机器人和人工智能伦理丛书

中国机器人
伦理标准化前瞻
(2019)

Chinese
Prospects for
the Standardization
of Robot Ethics 2019

北京大学
国家机器人标准化总体组————著

图书在版编目（CIP）数据

中国机器人伦理标准化前瞻.2019/北京大学国家机器人标准化总体组著.—北京：北京大学出版社，2019.10
（机器人和人工智能伦理丛书）
ISBN 978-7-301-30842-4

Ⅰ.①中… Ⅱ.①北… Ⅲ.①机器人–技术伦理学–标准化–中国– 2019 Ⅳ.①TP242-65 ②B82-057

中国版本图书馆 CIP 数据核字（2019）第 219441 号

书　　名	中国机器人伦理标准化前瞻（2019）
	ZHONGGUO JIQIREN LUNLI BIAOZHUNHUA QIANZHAN 2019
著作责任者	北京大学国家机器人标准化总体组　著
责任编辑	田　炜
标准书号	ISBN 978-7-301-30842-4
出版发行	北京大学出版社
地　　址	北京市海淀区成府路 205 号　100871
网　　址	http://www.pup.cn
电子信箱	pkuwsz@126.com
电　　话	邮购部 010-62752015　发行部 010-62750672
	编辑部 010-62750577
印 刷 者	北京中科印刷有限公司
经 销 者	新华书店
	880 毫米 × 1230 毫米　16 开本　11.25 印张　112 千字
	2019 年 10 月第 1 版　2019 年 10 月第 1 次印刷
定　　价	50.00 元

未经许可，不得以任何方式复制或抄袭本书之部分或全部内容。
版权所有，侵权必究
举报电话：010-62752024　电子信箱：fd@pup.pku.edu.cn
图书如有印装质量问题，请与出版部联系，电话：010-62756370

目录

缩略语 / 006

序言 / 008

执行总结 / 018

  致谢 / 018

  导言 / 020

  未来规划 / 028

第一章 机器人业与伦理 / 001

  1.1 人类的伦理生活 / 004

  1.2 机器人的现实伦理挑战 / 008

  1.3 BSI 和 IEEE 机器人伦理的经验与局限 / 014

第二章 中国机器人伦理宗旨 / 023

Contents

Abbreviations / 007

Foreword / 009

Executive Summary / 019

    Acknowledgements / 019

    Introduction / 021

    Looking Forward / 029

Chapter 1   Robotics and Ethics / 001

    1.1   The Human Ethical Life / 005

    1.2   The Actual Ethical Challenges of Robot / 009

    1.3   Lessons Learnt from BSI and IEEE / 015

Chapter 2   The Ethical Commitment of Chinese Robotics / 023

第三章　中国机器人伦理标准化体系 / 035

　　3.1　中国优化共生框架 / 038

　　3.2　中国优化共生设计方案的主要伦理目标 / 050

　　3.3　中国优化共生设计方案的实施方法 / 060

第四章　机器人伦理风险评估与应对 / 067

　　4.1　机器人伦理风险评估范围 / 070

　　4.2　机器人伦理风险评估表结构说明 / 072

　　4.3　机器人伦理风险评估表 / 078

第五章　护理机器人的伦理考量 / 093

　　5.1　护理机器人的界定 / 098

　　5.2　护理机器人的一般伦理原则 / 100

　　5.3　护理机器人的不同存在方式及其伦理后果 / 104

　　5.4　护理机器人伦理问题例示：智能假肢和机器宠物 / 114

参考文献 / 124

中国机器人伦理标准化前瞻委员会 / 138

Chapter 3　The Chinese System for the Standardization of Robot Ethics / 035

 3.1　The Fundamental Framework of the Chinese Optimizing Symbiosis Design Programme (COSDP) / 039

 3.2　The Principal Objectives / 051

 3.3　A Methodology for implementing the Chinese Optimizing Symbiosis Design Programme (COSDP) / 061

Chapter 4　Ethical Risk Assessment and Mitigation / 067

 4.1　The Scope of the Robotic Ethical Risk / 071

 4.2　Explication of the Table of the Robotic Ethical Risk / 073

 4.3　The Table of the Robotic Ethical Risk / 079

Chapter 5　Ethical Reflections on Care Robots / 093

 5.1　The Definition of Care Robots / 099

 5.2　General Ethical Principles for Care Robots / 101

 5.3　Different Modes of Care Robots and Their Ethical Implications / 105

 5.4　Artificial Prostheses and Robot Pets / 115

References / 125

Committee Members / 139

缩略语

BSI	英国标准委员会
COSDP	中国优化共生设计方案
EGE	欧盟科学与新技术伦理组
EURON	欧洲机器人学研究网
IEEE	电器与电子工程师协会
IFR	国际机器人学联合会
ISO	国际标准化组织
SAC	中国国家标准化管理委员会
UN	联合国

Abbreviations

BSI	British Standards Institution
COSDP	Chinese Optimizing Symbiosis Design Programme
EGE	European Group on Ethics in Science and New Technologies
EURON	European Robotics Research Network
IEEE	Institute of Electronical and Electronics Engineers
IFR	International Federation of Robotics
ISO	International Organization for Standardization
SAC	Standardization Administration of the People's Republic of China
UN	United Nations

序言

王天然
中国工程院院士
中国科学院沈阳自动化研究所

追随着机器人技术的进步和全球机器人产业的蓬勃发展，机器人的应用场景已经由早期的工业领域，扩展到了教育、医疗、家庭、安防等领域，机器人为社会带来了新的增长点，为家庭带来新的生活模式。

机器人与人工智能技术及其他科学技术的深度结合，使得机器人的智能化水平、适应环境的能力、与人交互和协作的能力不断提高。未来机器人会越来越多地融入人类的生活中，与人紧密协调，人和机器人会逐步形成新的"人机共融关系"。例如在制造业中，工人和机器人打破物理区隔，互相配合，共同完成一项作业任务；在家庭生活中，服务机器人与人存在于同一空间，机器人能够与人聊天、护理病人、完成家务等。这些都意味着机器人将从"工具"变为人类的"合作伙伴"。

Foreword

Wang Tianran
Academician of Chinese Academy of Engineering
Shenyang Institute of Automation
Chinese Academy of Sciences

Consistent with the rapid progress in robotic technologies and the global boom of the robotics industry, the role of robots in society is being extended from the familiar locus of industry to areas such as education, healthcare, family life, and security. Robots are adding energy to economic growth, and are producing a new mode of life in our homes.

The productive intersection of robotics, AI, and other technologies is raising the level of robotic intelligence, and facilitating their adaptation to the environment and their interaction with humans. In the future, robots will become increasingly integrated into human life and will collaborate closely with us. A truly "harmonious relationship" between humans and robots is taking shape. For example, we are lowering the physical barrier against robots and are working together with them on the factory floor. At home, service robots share our living space, are capable of keeping us company, provide home care for us, and undertake a multitude of tasks. Such developments show clearly that the role of robots is changing from being mere "instruments" to becoming our "cooperative friends."

机器人与人共融的紧密关系，给人类工作和生活带来极大便利，但也给我们原有的社会秩序带来了新的挑战。比如工业机器人与人共同工作，一旦发生事故，责任如何划分；家用服务机器人进入家庭，会不会泄露个人隐私，人类会不会对其产生"感情"；服务于医院的公共机器人决策过程中会不会带有偏见，例如分配急诊病人的优先次序时带有偏见等。人们担忧机器人不知何时会失控、在人类不知情的情况下侵犯人类的权利，更担忧这些行为无法追溯和问责。

这些问题并非是技术瓶颈，而是机器人如何"做决定"的问题，也就是机器人伦理问题。这些问题的解决方式直接决定了公众对于机器人及其技术的接受程度。因此如何对机器人的行为加以约束，如何引导机器人的行为和公共决策能够符合人类的基本价值观，成为机器人设计者、生产者、应用者乃至社会管理者都十分关注的课题。本书以此为出发点，从标准化的角度，来研究和探讨机器人的伦理问题，提出了一整套机器人伦理体系，用于引导机器人的设计、生产和应用，规避机器人的伦理风险。

我深信，本书必将引起社会的广泛关注，引发全社会对于机器人伦理的热烈讨论，吸引更多的有识之士投入到机器人的伦理研究中。同时希望这项研究能够为未来机器人伦理标准的制定乃至伦理法律治理奠定基础，推动机器人的设计和研发越来越有益于人类和自然的和谐发展。

序言

While robots offer significant benefits in the workplace and in other life activities, at the same time the intimate relationship emerging between robots and human beings also challenges our existing societal order. For example, how to assign liability becomes unclear when accidents occur in the collaboration between industrial robots and humans. Again, service robots raise the risk of leaking private data when they enter our home space. It has also been discovered that human emotions function to bind us unilaterally to our robot friends. Furthermore, healthcare robots in the hospital environment could conceivably manifest biases in the processes of decision-making such as prioritizing certain groups of people in emergency services. And finally, there is real concern that we might lose control over robots to the extent of human rights being violated, and the difficulty of accountability and assigning liability as such cases occur.

Such problems are not technical in nature, but rather raise concerns about how robots "make decisions," and thus are problems of robot ethics. Success in solving such problems will have an immediate impact on the public's willingness to accept robots and robotic technologies. It is for this reason that how to restrain robotic "behaviours" and how to align robotic "behaviours" and "decisions" with our fundamental values becomes a major concern of robotic designers, manufacturers, users, and government agencies. Facing such challenging issues, this publication is a strong beginning in the research and discussion necessary to bring standardization to robot ethics, and offers a comprehensive ethical system to preempt ethical risks and hazards in the expanded use of robots and that can be used to orient efforts in robotic design, manufacture, and application.

I am certain that this present work will attract wide attention among the public, and stimulate robust discussions concerning robot ethics across the whole of society, and in so doing, will encourage more people to undertake research on robot ethics. At the same time, I am sure that this work will serve to ground our ethical standards and inform our legal governance over robotics for the immediate future, and will inspire the kind of robotic design and development that will foster a harmonious relationship between humans and nature.

王博

哲学系教授

北京大学副校长

从来没有任何一种生物像人类一样试图通过知识和技术改变这个世界，并取得了巨大的成功。这种成功对于人类生活的影响显而易见。从平均寿命的延长到日常生活的便利，从外太空的探索到精神世界的拓展，人类不断地赞美和惊叹于自己的力量，同时思考这种力量是否会让人类的未来充满了更多的不确定性。机器人革命时代的来临让人们更加兴奋，也让人们更加忧虑。

对于技术的担心贯穿人类精神生活的始终。古代中国的哲人庄子曾经说过："有机械者必有机事，有机事者必有机心。"在直接的意义上，庄子忧虑技术给人心和人类生活带来的不可逆的负面影响。而在更深刻的意义上，他意识到的是知识和技术的有限性，它们固然可以解决一些问题，却也带来了更大的问题。《周易》同样指出单向度的"进"会导致"亢龙有悔"的局面，肯定"知进退存亡而不失其正者"，才是所谓的圣人。庄子和《周易》的思考当然无意也无法阻挡知识和技术的进步，但仍然有助于人们在热闹中保持清醒的头脑，兼顾科技和人文的平衡。

序言

Wang Bo

Professor of Philosophy,

Vice-President of Peking University

There has never been a species like humankind that has attempted to change its world through knowledge and technology, and that has been so successful in doing so. The impacts of this knowledge and technology on our way of life is everywhere evident. From a much-extended longevity and the conveniences of everyday life to space exploration and expanding spiritual world, we human beings have accomplished much and have come to extol our own power in accruing such magnificent achievements. At the same time, we cannot help but wonder whether or not the unleashing of such power will bring with it real uncertainties for the human experience. The pending era of robotic revolution excites and worries us both at once.

Concern over the impact of new technologies is pervasive in human life. Zhuangzi, one of the most prominent sages in ancient China, once stated: "where there are machines, there are bound to be machine worries; where there are machine worries, there are bound to be machine hearts". On the surface, Zhuangzi seems to be concerned about the irreversibility of the negative impacts that technology may have on the human life and spirit. Yet at a deeper level, he offers us a profound reflection on the limits of our knowledge and technology. For him, knowledge and technology create more problems than they solve. The *Yijing*, as the first among the Chinese classics, avers that progress might bring with it the disastrous consequences captured in the image of "the wilful dragon repents" and says further that only those persons who "while understanding progress and recession, life and death, can still maintain their integrity" can be called a sage. Neither of these canonical texts are recommending that we abandon intellectual and technological progress, but rather that in our enthusiasm to move ahead we remain clear-headed and maintain a balance between scientific knowledge and our humanistic concerns.

中国传统艺术的一个重要观念是留白。基于对有和无、实和虚之间关系的理解，艺术家们并不想占有全部的空间，而是把很大的一部分留给事物，也留给欣赏者。这不仅创造了美，也创造了事物的自由流动和欣赏者自由的想象。在这个意义上，留白是善，也是智慧。对于机器人、基因工程等技术来说，意识到留白的妙处，是非常必要的。

哲学家的特点是从整体上来理解世界，关注事物的多个面向，并寻求最恰当的生活。这个特点体现在《中国机器人伦理标准化前瞻（2019）》（以下简称《前瞻》）的工作中。以整体论的善为根基，围绕着多元、自然、正义和繁荣，倡导世界共生秩序，使这项工作充满了对伦理价值的尊重和社会责任的担当。而去人类中心主义的立意不仅让《前瞻》关心人类福祉，也关心自然存在物，引导机器人业通向一个共生和谐的良性世界秩序。我们可以看到这项工作的主持者们自觉地吸取传统中国的思想资源和现代世界的一般价值，力图使这份中国优化共生设计方案取得广泛的国际共识。希望这个方案可以引起科学家、人文学者、企业家、政治家和普通人群的关注，在广泛的讨论中得到进一步的完善。

In Chinese traditional painting, one of the most important concepts is to leave spaces empty. Based on their understanding of the relation between presence and absence and between fullness and emptiness, Chinese painters never occupy the entire space of a picture but leave most of it blank for its figures and for their observers. The art of leaving blank space not only expresses a kind of beauty, but also allows for the flow of the figures and for the free play of imagination on the part of the observers. In this sense, leaving blank space is felicitous, and manifests a kind of wisdom as well. On the same principle, it is thus indispensable to acknowledge the significance of leaving blank space in developing cutting-edge technologies of robotics and gene editing.

An important characteristic of Chinese philosophy is to understand the world holistically, and in thus seeing things from multitude of perspectives, to seek after the most appropriate form of life for all of us. Such an attitude is much in evidence in the production of the *Chinese Prospects for the Standardization of Robot Ethics 2019* in its respect for ethical values and social responsibility. The *Prospects* are rooted in a holistic concept of an optimal good from which it derives a formal framework that includes the fundamental principles of pluralism, nature, justice and well-doing. With its commitment to optimizing symbiosis, the *Prospects* eschew any kind of anthropocentrism in its approach to human welfare and the protection of the natural environment, and seek to incorporate the development of robotics into a well-ordered and integrated world. The authors of the *Prospects* have deliberately drawn upon both the Chinese traditions and modern values in developing Chinese Optimizing Symbiosis Design Program (COSDP) that has the capaciousness to attract a broad international consensus. It is our hope that the proposed program will draw the attention among scientists, humanities scholars, entrepreneurs, decision-makers and the public at large, and that it can be further improved upon by continuing discussion among all of the constituents of society.

北京大学一直关注变化世界中人类所面临的重大问题和挑战。学校在人文、社会科学和科学等领域的知识和人才储备，无疑是主导这项工作的基础。而不同学科和机构学者之间的合作，让这项工作的意义超出了工作本身。人文学者对于科技革命以及它所改变的世界的关注，会让人文学更具生命力和影响力。而科学家们的人文情怀和视角，也会让智能科技时代充满对事物和人类体贴的精神。

序言

Peking University is unrelentingly attentive to the big questions and challenges that we face in an always changing world. Our rich intellectual sources in the areas of the humanities and in the social and natural sciences have provided a strong and certain foundation for producing the *Prospects*. The kind of cross-disciplinary collaboration among scholars from various institutes and disciplines makes such an undertaking more significant than the assemblage of any particular product. For the humanists, attention to the revolution in the sciences and to ongoing advances in technology can instill vigor and extend the reach and influence of our humanities research. And for the scientists, an awareness of the perspective brought by the humanists in this age of artificial intelligence will contribute importantly to appropriate concerns about both our world of things and our own species.

执行总结

致谢

在中国国家标准化管理委员会的精心组织和协调下,国家机器人标准化总体组与北京大学哲学系和外国哲学研究所紧密合作,正式制定并颁布首部《中国机器人伦理标准化前瞻(2019)》。

《中国机器人伦理标准化前瞻(2019)》的成功撰写依赖众多机构、学者和机器人业参与者的支持和帮助。我们首先诚挚感谢中国国家标准化管理委员会和国家机器人标准化总体组远见卓识地设立引导专项,及时启动研究并制定中国机器人伦理标准化体系。我们感谢国家机器人标准化总体组对机器人伦理标准化工作始终不渝的信任、鼓励和支持。

我们感谢北京大学,北京大学哲学系和北京大学外国哲学研究所从始至终对这部《中国机器人伦理标准化前瞻(2019)》研究和制定工作的鼓励与支持。

这份文件的撰写来自北京大学和浙江大学专业学者敬业、默契、合作和高效的集体工作。我们感谢工作组所有成员的忘我付出、出色行文、严谨分析和睿智思路。

Executive Summary

Acknowledgements

Under the guidance of the Standardization Administration of the People's Republic of China, The Department of Philosophy and the Institute of Foreign Philosophy, Peking University, have collaborated with the National Robotics Standardization Committee to produce this document *Chinese Prospects for the Standardization of Robot Ethics (2019)*.

We would like to thank the Standardization Administration of the People's Republic of China (SAC) and the National Robotics Standardization General Working Group of China for taking the timely initiative to elaborate a Chinese robot ethical standard. We again thank the National Robotics Standardization General Working Group, for the persistent trust, encouragement and their support.

We would also like to thank the Department of Philosophy and the Institute of Foreign Philosophy at Peking University for their continuing encouragement and support for this *Prospects* project.

This document is the result of cooperative team work of academic people from both Peking University and ZheJiang University. We thank all the members of the team for their energetic engagement with this project, for their substantial contributions to it, for the rigor they brought to the analysis, and for the intelligent solutions they had on offer.

导言

机器人革命时代正在来临！进入21世纪以来，得力于汽车制造以及电子电器工业的拉动，工业机器人生产显著提速。与此同时，服务机器人在物流、国防、作业、医疗、家居以及教育娱乐领域开始崭露头角，正在为人类社会和家庭带来新的生活模式和生活前景。

随着全球经济、科技竞争加剧以及人口老龄化问题日益凸显，世界各主要机器人研发国家（美国、欧盟、英国、日本和韩国）都相继制定了自己的机器人发展规划。中国政府也早在2006年就开始部署自己的机器人发展战略。2012年以来，中国政府陆续出台各项具体政策和指导意见，加速推进中国机器人业的发展。如今，机器人业已被政府列为着重发展的十大重点领域之一。

机器人业的蓬勃发展毫无疑问将直接冲击每一个人的权利和福祉，对人类社会产生广泛而深远的影响。20世纪40年代，著名科幻作家阿西莫夫（Isaac Asimov）就在自己的科幻小说中思考机器人生产设计以及机器人自身行为的伦理准则，并深有远见地提出了自己的"机器人学三定律"。如果阿西莫夫三定律还只是以科幻方式来设想机器人的潜在风险，今天我们已经不可避免地遭遇现实机器人造成的现实风险和现实挑战。为此，全球主要机器人研发国家和地区以及国际相关专业机构都开始采取积极行动，制定应对策略，迎接机器人业可能带来的社会和伦理挑战。

Introduction

The robot revolution is coming. Driven by the automotive and the electrical and electronics industry, the demand for industrial robots has accelerated markedly in the first two decades of the twenty-first century. In the past years, service robots have started to be deployed in such areas as logistics, defence, agriculture, surgery and medical procedures, healthcare, household and "edu-tainment". The increased use of robots is transforming the individual and the social life of everyone.

Because of the global competition in manufacturing economics and the rapid increase of their aging population, the USA, UK, EU, Japan and the Republic of Korea as the most prominent countries in the development of robotics, one after another, set up each of their strategic plans to encourage and develop robotic technologies. By 2006, Chinese government had already initiated a national strategy for Chinese robotics, and since 2012, more national policies and guidelines were introduced to accelerate the development of Chinese robotics. In 2015, robotics was included as one of the ten strategic industries in the country.

This unprecedented flourishing of robotics has not only made an enormous impact on the individual rights and the well-doing of human beings, but also on industrial production and social life. As early as the 1940s, Isaac Asimov, a famous science fiction novelist, began reflecting on ethical rules for the robotic design and the 'autonomous' behaviour of robots themselves in his writings. As a result, he with an incredible prescience worked out what has become well-known as "Asimov's Three Laws of Robotics". While robotic risks were fictitious and mere speculation at the time of Asimov, today with the development of robotic technologies we have already become entangled in numerous unavoidable hazards, risks and injuries. An adequate policy framework must be put in place to allow China to be at the forefront of the design and deployment of robotic technologies while at the same time avoiding safety risks for people in their homes and workplaces. All of the most prominent countries and regions, and professional institutes of robotics have undertaken initiatives to elaborate an implementable strategies to respond to the ethical challenges that arise from the development of robotics.

1. 目的

面对机器人业给人类带来的前所未有的伦理挑战，作为世界机器人技术的重要研发国家之一的中国同样责无旁贷。《中国机器人伦理标准化前瞻（2019）》的制定与颁布旨在为未来机器人业相关科技领域的研发提供洞见和建议。

《中国机器人伦理标准化前瞻（2019）》的核心目的在于，**表达中国政府和中国公众对于机器人业发展的深刻伦理关切，推动并培育全社会关于机器人的理性认知和健康讨论，为未来机器人业的伦理指南和细则制定开辟道路，推动中国机器人业产学研生态结构的调整升级**。通过制定和颁布这部《前瞻》，我们希望推动机器人的设计和研发向着与伦理价值相一致的方向发展，在不断推进机器人科技创新和产业发展的同时，使中国机器人业的所有参与者同样具有一颗充满伦理关怀和社会责任的人文之心。

2. 中国机器人伦理标准化体系

在世界各国，机器人业的长足发展都离不开对伦理价值的尊重和对社会责任的担当。《中国机器人伦理标准化前瞻（2019）》所致力于提供的是中国机器人伦理标准化体系。这份文件并不是要成为中国机器人业科技创新的桎梏，而是要推动伦理价值以前瞻性方式呈现在未来机器人的设计之中。《前瞻》提供的中国机器人伦理标准化体系既植根于中国多元的文化传统资源，也广泛吸纳现代人类价值和其他文化的伦理观念。这套标准化体系由中国优化共生设计方

1. Mission

While aspiring for a leading role in the core technologies of robotics, China must never neglect its critical ethical responsibility for the Chinese robotic design and deployment. This document is meant to provide insights and recommendations for the technological advance relevant to robotics in the coming years.

The *Prospects* are committed *to stating the deep ethical concerns of the Chinese government and the public with respect to robotics, to cultivating and promoting rational discussion and understanding in the entire society, to paving the way for the future elaboration of ethical guidelines and regulations in robotics, and finally to improving and upgrading the ecosystem of Chinese robotic research, education and industrial production.*

In this way, this document seeks to promote the design and deployment of robotic technologies in alignment with human ethical values. While continuing to make its contributions to technological innovation and industrial growth, Chinese robotics will not abdicate its unquestionable social responsibilities and its ethical concern for humanity.

2. The Chinese System for the Standardization of Robot Ethics

In the world, the development of robotic technologies cannot be sustained without an adequate awareness of and respect for human ethical values and social responsibilities. This document provides the fundamental system for the standardization of ethical rules governing Chinese robotics. This document does not intend to hinder or arrest technological innovations and the development of Chinese robotics but to in a prospective way promote the possible embodiment of human ethical values in every robotic design. The Chinese system for the standardization of robot ethics consists of multiple layers within which the fundamental framework can be maintained as persistent and stable foundation. The Chinese system is rooted in the values of the Chinese classical traditions, and at the same time, seeks as broadly as possible to incorporate modern human values and the ethical ideas of other communities. The Chinese system for the standardization of robot ethics consists of the Chinese optimizing symbiosis de-

案（COSDP）和相应的实施方法共同构成。它的核心目标在于通过卓越的机器人技术设计和研发来**优化世界共生秩序**。

我们所倡导的世界共生秩序要求对善提出一种整体论的理解。根据整体性的逻辑要求以及人类生活的基础经验，中国机器人伦理的整体善必须同时展现为四个元向度（多元、自然、正义和繁荣）。由于人类逻辑思考和基础生存经验的普遍性特征，中国机器人伦理标准化体系能够在世界各个文化共同体中赢得最广泛的重叠共识。依据当前国际社会关于机器人各项伦理挑战的主流意见，中国机器人伦理标准化体系的四个元向度又进一步具体化为涵盖机器人设计、运行和使用过程的五个主要伦理目标：

1. **人的尊严与人权**：根据中国政府承认的国际人权公约以及其他国际专业机构文件，国家必须通过规范、政策、研究和执法来确保所设计的机器人系统不侵犯人的尊严和人权。

2. **责任**：确保机器人的设计者和操作者能够明确责任归属并承担责任。

3. **透明度**：在机器人的检测、使用和事故调查中，确保系统以透明和在所要求范围可被解释的方式运行。

4. **避免滥用**：培育全社会以及国家政府和司法机构充分意识到机器人滥用带来的严重危害，以有效避免或减少机器人滥用。

5. **共生繁荣**：机器人业的发展要全面考虑对人类生活质量的影响。机器人设计和研发不仅要尊重各个文化传统价值的差异性，而且要尊重超出人类控制和利用的自然环境与自然物种的内在价值。

sign programme (COSDP) and a methodology for its implementation. The principal purpose of the programme is to *optimize a symbiotic world order* through an outstanding design and deployment of robotics.

This idea of a symbiotic world order entails a holistic conception of the good. Acknowledging the logical implication of ethical holism and our respect for biodiversity, the Chinese robot ethics is to be responsive simultaneously to the following four meta-dimensions: pluralism, nature, justice and well-doing. Given the general nature of human rational thinking and the fact of biodiversity, the Chinese system for the standardization of robot ethics is designed to achieve an overlapping consensus among diverse cultural communities all over the world.

The Chinese system for optimizing a symbiotic world order is particularized in the context of today's global consensus on the ethical risks of robotics. The four meta-dimensions delineated above are thus stipulated as the following five major objectives that, as ethical criteria, are needed to orient robotic design, operation and use.

1. *Human dignity and human rights*: according to the international documents produced by the UN and professional institutes on human rights, stakeholders should ensure that robotic systems do not compromise human dignity or violate human rights in their regulations, policies, oversight and enforcement.

2. *Accountability*: robotic system should be designed to ensure the possibility of tracking responsibilities to their designers, operators and users and hold each of them accountable.

3. *Transparency*: robotic systems in their use and oversight should be accessible and opened to the verification and validation.

4. *Awareness of misuse*: stakeholders should educate and promote the awareness of the entire society and especially the government as to the hazards of robotic misuse in order to minimize if not avoid the risks of misuse.

5. *Prioritizing shared flourishing*: stakeholders should take full account of the possible impact that robotics might have on the quality of life. The design and deployment of robots should not only respect cultural diversity, but also

中国优化共生设计方案（COSDP）正是由上述整体论含义的善、四个元向度以及五个具体伦理目标构成的三层伦理系统。未来，与伦理相一致的机器人设计和研发就意味着中国机器人业要遵循中国优化共生设计方案。

中国优化共生设计方案需要建立自己独特、有效的实施方法。中国机器人业要通过伦理主导的跨学科研究、开放的讨论协商以及反思平衡式的伦理分析来施行该方案。借助中国机器人伦理标准化体系，未来中国机器人业不仅能够鉴别、减轻、阻止和解释机器人使用对社会和个人生活造成的不良影响，而且能够制定机器人伦理标准，并把人类生活和自然世界的内在价值秩序通过机器人技术完美体现。

3. 内容概览

不同于BSI，IEEE以及欧盟委员会三大机构公布的机器人相关伦理文件，《中国机器人伦理标准化前瞻（2019）》不再把人类中心主义作为自己伦理标准化体系的根本理论依据。中国机器人伦理标准化体系所致力的是具有整体特征的美好共生秩序。

这份文件的第一章将展示中国机器人伦理标准化系统提出的理论和现实背景。我们将在现实存在的机器人伦理问题语境中来展示人类中心主义框架的缺陷。第二章将申明中国机器人伦理标准化的独特路径。第三章将阐释中国共生优化设计方案，并提出中国机器人伦理标准化的主要伦理目标和实施方法。第四章要建立机器人的

the intrinsic values of the natural environment and non-human entities that lie beyond human control and exploitation.

The entire system for the standardization of robot ethics thus has three layers: the shared good of the symbiotic world order, the four meta-dimensions and the five principal ethical objectives. The proposed ethically aligned design and deployment of robots seeks to implement the Chinese optimizing symbiosis design programme in robotics.

The Chinese optimizing symbiosis design programme can only be implemented through a combination of cross-disciplinary research, open discussion and reflectively balanced ethical deliberation altogether. In according with the Chinese programme, stakeholders will be able to not only identify, mitigate, arrest, and interpret any unintended negative consequences of robots on lives of individuals and society, but also promote the embodiment of the intrinsic values of human life and the natural world in robotic technologies.

3. Survey of the Contents

In contrast to other published robot ethical documents and standards set by other countries or professional institutes, the *Prospects* do not accept the prevailing anthropocentric principle any more. Chinese robot ethics is committed to a well-ordered symbiosis in the holistic sense stated above.

In the first chapter, this document will present both the theoretical and historical background of the Chinese optimizing symbiosis design programme for robot ethics. The second chapter will state the unique features of the Chinese approach to robot ethics. The third chapter will then outline the Chinese optimizing symbiosis design programme, and explain the principal ethical objectives and its method of implementation. The fourth chapter will elaborate the basic system of robotic ethical risks and the principles of

伦理风险和应对原则的基础系统。近年来，护理服务机器人在中国和其他国家市场上呈现出巨大商业潜力。《前瞻》的最后一章将通过护理服务机器人的具体案例来展示中国优化共生设计方案的可行性前景。需要说明的是，鉴于中英文语境的差异，这份文件的中文和英文文字并不是两个文本之间简单和直接的对译。

未来规划

《中国机器人伦理标准化前瞻（2019）》先导工作为中国机器人业制定了由中国优化共生设计方案和实施方法构成的伦理标准化体系。这套机器人伦理标准化体系不仅是中国机器人业和公众的一份郑重的伦理声明，而且需要在未来对既有机器人系统改进和对新技术创新中被严格考虑和执行。由于机器人技术方兴未艾，中国优化共生设计方案，特别是目前它所包含的主要伦理目标，在可见的未来将会得到一定调整。这部前瞻文件对于未来执行该伦理标准化体系时的必要修正保持开放。

中国优化共生设计方案应该通过系统性制定机器人伦理标准、导则或规范来获得具体化和实际执行。一些标准普遍适用所有类型的机器人，而其他导则必须针对不同类型机器人来分别制定。今天，中国机器人业根本无法封闭式地独立发展，而必须融入国际合作和全球市场。只有充分考虑世界文化的多元特征，中国机器人设计与应用的伦理导则才能得以妥善制定和良好施行。有益人类的机器人

mitigation. Since healthcare robots show a considerable growth potential both in China and other countries, the final chapter of the document will present a case study on the ethical risks of healthcare robots to illustrate the prospects for implementing the Chinese optimizing symbiosis design programme. Because of a remarkable difference of cultural context, it is preferable to avoid a simple and literary translation between Chinese and English version of the document.

Looking Forward

The Chinese *Prospects* initiative has created the Chinese optimizing symbiosis design programme (COSDP) as the Chinese system for the standardization of robot ethics. This programme is not merely an ethical statement on the part of Chinese robotic and the public society but also needs to be seriously considered and implemented in the improvement of existing robotic systems and the innovation of new technologies. Because of the emerging nature of robotic technologies, the programme, particularly the principal objectives of the programme, will likely be adjusted in the near future. The *Prospects* will be kept open to the necessary revision on the basis of the future implementation of the programme.

The Chinese optimizing symbiosis design programme should be implemented and concretized through particular guidelines or regulations. Some of the guidelines are common to all robots whereas the others must be differentiated according to types of robots. Today Chinese robotics industry cannot separately develop but be immersed in international collaboration and global market. The ethical guidelines of robotic design and deployment will only be very well developed when taking account of cultural pluralism. The border for a beneficial robotics may often look very obscure and

业边界往往比较模糊，甚至充满争议。我们只有依赖严格而融贯的哲学论证，同时借助跨学科的理论支持，才能塑造具有可操作性的社会共识并以伦理的方式引导未来机器人的设计和应用。为此，在国家层面设立专业性机器人伦理监管委员会势在必行。未来，国家机器人伦理监管委员会将负责不同类型机器人伦理标准的制定，并为不同机器人技术试验颁发伦理执照并执行相关检查。

《中国机器人伦理标准化前瞻（2019）》先导工作希望把中国机器人伦理标准化体系和未来的导则或规范最终转变为教育、教学材料。这些教育材料将会引发学术界、工程师、技术人员、博士后、社会媒体以及公众的浓厚兴趣和广泛关注。我们相信，在今天这个科技文明的时代，公众广泛参与机器人业的伦理讨论将极大惠及未来中国机器人业的繁荣发展。

controversial. Only a philosophical rigor and consistency with the aid of cross-disciplinary contributions can be sufficient to structure practical consensus to ethically orient the future design and deployment of robots. A special national committee for robot ethics thus seems to be necessary to elaborate ethical guidelines or regulations for robotics and to license for and inspect experiments of disruptive technologies in future.

The *Prospects* initiative would like to eventually convert both the framework and the future guidelines or regulations into educational materials. These would interest academics, engineers, technologists, post-graduates, social mediate and the public. The wide involvement of the public into the ethical discussion of robotics will immensely benefit the flourishing of Chinese robotics.

第一章　机器人业与伦理
Chapter 1　Robotics and Ethics

世界机器人业的发展已逾半个世纪，中国早已成为全球最大的工业机器人市场。今天专业服务机器人和个人或家用机器人也迎来了蓬勃发展的曙光。机器人业正在为各国经济增长、人类生活质量提升以及人类能力增强做出自己的巨大贡献。

历史上，新的科技发明总会带来世界整体图景的深刻转变并广泛重塑人类社会的生产生活模式。在充分享受机器人带来的各项福祉的同时，我们急需仔细分析并审慎应对机器人对个人以及社会生活造成的巨大冲击和影响。今天，机器人带来的伦理挑战不再是科幻电影和文学作品玄想的主题，而已然成为我们必须面对的严峻现实。国际上机器人主要研发国家和地区（美国、英国、欧盟、日本和韩国）都无一例外地开始采取主动措施，制定机器人相关的伦理框架、指南或标准。EURON最早发起并发布了《机器人伦理路线图》(2006)，英国BSI出版了《机器人和机器人设备：机器人及机器人系统伦理设计和使用指南》(2016)，美国IEEE建立专门的伦理和法律委员会并组织全球300余名专家先后撰写和发布了两版《与伦理协调的设计：自主和智能系统中人类福祉优先远景》(2016 & 2017)，欧盟议会正式通过了欧盟委员会提交的《机器人学民法条例》(附机器人伦理文件)报告（2017）[1]，欧盟EGE出版了《关于人工智能，

第一章 机器人业与伦理

The world robotics industry has been developing for more than half a century, and China has become the biggest industrial robots market all over the world. Today both professional and personal/domestic robot sales are soaring. Robotics is making a significant contribution to economic growth, the improvement of the quality of human life, as well as the empowerment of our human capacities.

In history, new technologies have always brought forth changes in our worldviews and have in the fullness of time affected the moral lives of human beings. While taking full advantage of all of the unprecedented benefits robots provide, people are urged to think through and find ways to cope with the enormous effects that the robotics industry is likely to have on individuals and society. Today the ethical challenges that robots occasion are no longer the stuff of science fiction and idle speculation, but have become our reality. The most prominent countries in technological innovations and the development of robotics (such as the EU, US, UK, Japan and the Republic of Korea) early on made known their intentions to produce and publicize their ethical and legal guidelines, and their frameworks and standards regarding robotics. EURON *Roboethics Roadmap* (2006), BSI *Robots and Robotic Devices* (2016), IEEE *Ethically Aligned Design* (2016 & 2017), European Parliament *Report with recommendation to the Commission on Civil Law Rules on Robotics* (2017)[1], EGE *Statement on Artificial Intelligence, Robotics and 'Autonomous' Systems* (2018)[2] and The European Commission's High-Level Expert Group on

机器人学与自主系统的声明》(2018)², 欧盟委员会人工智能高层专家组发布了征求意见稿《可信赖人工智能伦理导则》(2018)³。这些都是世界机器人业发展中具有远见卓识和社会责任担当的系统性工程。为了促进中国机器人业健康发展并提升国际竞争力，中国在机器人伦理规范制定工作中同样责无旁贷。

巨大的机器人市场和广泛多样的机器人需求构成了未来中国机器人业快速发展的决定性驱动力量。然而，经济和财政繁荣绝不是中国机器人业的唯一关注。对于每一位涉足中国机器人业的参与者而言，自然和文化的宝贵价值同样具有无法替代的重要意义。这一章要展示机器人业发展与人类社会和自然世界内在价值的密切关联。为此，我们不仅要描述人类生活的伦理特征，而且要展示机器人对人类社会已经造成的伦理冲击和挑战。面对棘手的挑战，BSI 和 IEEE 制定的先驱性文件为中国机器人伦理标准化工作提供了广泛参考和借鉴。然而，他们的理论框架和具体规范仍然可能在未来遭遇难以应对的伦理困难。在本章中，我们将从中国传统伦理视角出发对这些潜在困难进行分析。

1.1　人类的伦理生活

人类伦理是人们为了生存、繁衍、发展和繁荣而在社会生活中发展出来的一套关于应当如何生活和行动的思想观念，并对人们在社会生活中所能采取的行为加以规范和约束。一般来说，伦理规范具有两个本质特点：第一，它们所表达的价值（即伦理价值）与社

Artificial Intelligence *Draft Ethics Guidelines for Trustworthy AI* (2018)[3] are all timely and systematic publications on robot ethics. For the goal of robust development in this industry, Chinese robotics cannot afford to neglect the robot ethics that must evolve with it.

The huge Chinese market and high demand are decisive forces that will drive the rapid advance of Chinese robotics. But the economic and financial prosperity can never be its single concern. The precious values of nature and culture must be of equal importance for all of the Chinese stakeholders involved in robotics. This section clearly presents the implication of intrinsic values of natural world and human society in the development of robotics industry. For this purpose, we not only describe the ethical nature of human life but also indicate ethical consequences and challenges which the development of robotics has made on human society. As pioneering replies to ethical challenges of robots, the BSI and IEEE documents certainly serve as valuable references for us. At the same time, their ethical frameworks and regulations may still leave certain ethical problems unresolvable in future. In this chapter, the implicit limitations of their standards and documents are reviewed and analyzed from the perspective of the Chinese classical traditions.

1.1 The Human Ethical Life

Ethics is a systematic set of ideas that human beings develop in order to survive, reproduce, advance, and flourish in society, and it impose normative constraints on what persons can do with respect to others within the context of the human social life. Generally speaking, ethical norms have two essential features: first, the values they are supposed to reflect are intimately related to the basic conditions for social cooperation, harmony and flourishing, and are deemed to be of critical importance and to have some degree of universality;

会合作、社会和谐以及社会繁荣的基本条件具有本质联系，在人类生活中具有根本重要性和某种程度的普遍性；第二，它们对人们提出的要求往往通过内在良知和伦理情感来实现，与人们的认知能力和情感能力具有重要联系。

伦理规范必须具有引导人们行动和选择的基本职能。因此，伦理规范对人们提出的要求往往需要明确的规则或原则来表达。鉴于人类生活的复杂性、伦理价值与其他价值的错综复杂联系以及多元的伦理传统，伦理规范不可能总是采取规则或原则的形式。在伦理规则或者它们所规定的义务之间发生冲突时，或者在严格意义上的伦理价值与其他价值发生冲突时，人们为了解决冲突，就需要设想某个高层次的伦理原则或理论。然而这样的原则或理论并非次次都能找到。即便可以提出某个原则或理论来解决特定伦理难题，相关的原则或理论也未必能够获得一致认同。伦理难题的存在不仅是对人类经验和实践智慧的普遍考验，而且会对机器人伦理提出严峻的挑战。

人类伦理生活要求复杂的精神能力。伦理能力的产生至少要求五种其他基本能力[4]：第一，情感和情感表达能力；第二，明确表达社会规则的能力；第三，将伦理情感或评价态度转移到行为本身的能力；第四，对伦理行为的发生进行追踪和记忆的能力；第五，把精神状态赋予其他行动者的能力。此外，伦理判断和伦理动机还提出了其他方面的能力要求。在可预见的未来，机器人尚无法具备人类伦理生活所要求的全部能力。基于现有机器人技术已然可能带来

second, the requirements they stipulate are normally satisfied by means of inner conscience and ethical sentiments, and thus have important implications for human cognitive and affective capabilities.

Ethical norms must be capable of guiding the choices and actions of human agents. Accordingly, the requirements which they put forward normally need to be expressed as explicit rules or principles. However, owing to the complexity of the human life and the intricate relations between ethical values and other competing value as well as the existence of a plurality of ethical traditions, ethical norms cannot always be expressed in the form of a rule or principle. When ethical rules or the duties stipulated by them conflict, or when ethical values are in conflict with other values, some higher-level moral principles or theories need to be worked out to resolve the tension. However, it may be the case that such principles or theories are not available, or else the principle or theory that is proposed for addressing a particular ethical challenge may not be endorsed unanimously by all involved agents. The existence of ethical dilemmas in real human life do not only constitute a general test for the human experience and its wisdom, but also produces a challenge of the most serious kind for the ethical design of robots as well.

A series of complex mental capabilities is required for a human ethical life. The emergence of ethical capabilities need, at a minimum, five other basic abilities[4]: 1) a range of emotions and the ability to express them; 2) the ability to formulate social rules; 3) the ability to transfer ethical emotions or evaluative attitudes to behaviour itself; 4) the ability to trace and remember the genesis of ethically significant actions; 5) the ability to attribute mental states to others. In addition, some other abilities are also required for human agents to make ethical judgments and to have ethical motivation. It may well be that robots will not have possessed all of the abilities necessary for leading a human ethical life

的隐患，我们目前最好首先把机器人伦理理解为关于设计、生产和使用机器人的人员的伦理。

1.2　机器人的现实伦理挑战

世界机器人业蓬勃发展的速度远远超出人们的想象。现在，各类机器人已经进入人们的社会生活，广泛用于劳动服务、国防安全、科研教育、医疗保健、个人陪护、环境保护、教育娱乐等各个领域。机器人不仅能够大大提高企业生产效率，减轻工人负担和风险，而且能够为人类日常生活带来诸多便利。机器人不再只是单纯为人所用的工具，而且也在社会交往中发挥一定作用，甚至会成为人类的朋友。然而，随着机器人进入人类生活，许多现实的社会和伦理问题也随之涌现。

20 世纪，核物理、生物技术以及计算机科技等尖端科技产业曾引发严重的伦理后果。例如，曾经如日中天的计算机产业不得不面对许多未曾预料的社会伦理挑战：企业裁员、隐私泄露、侵犯知识产权、现实世界异化、网络安全、网络霸凌、电信诈骗、网瘾等。[5]由于一些高科技产业酿成的恶劣影响，许多国家曾饱受公共舆论的激烈谴责和抨击，不得不终止相关技术研究和应用或者采取严格措施加以管控。[6]这些前车之鉴为今天机器人业的可持续发展提供了宝贵的经验教训。随着机器人业研发制造的日益壮大，人机互动更趋频繁，机器人业必然遭遇比核物理和计算机科技更为复杂的社会和伦理挑战。事实上，有些风险和问题已经造成了不良后果；还有其

in the foreseeable future. Thus, at present, it is best to regard robot ethics, first and foremost, as such a discipline that examines the ethical issues of human beings involved in the design, producing and use of robots.

1.2 The Actual Ethical Challenges of Robot

The robotic technologies and industries are evolving globally at an unimaginable pace. Today different kinds of robots are appearing everywhere in human society: in factory manufacturing, in domestic service, in scientific research, in education, in healthcare, as companions, in entertainment, in environmental protection, and so on. Industrial robots not only can increase productive efficiency dramatically but also alleviate labour demands and risks to human workers. Service robots can provide great benefits to human life. Because of their increasing intelligence and autonomy, robots cannot be simply regarded as mere instruments available to serve human purposes. Inasmuch as humans may even become friends with robots, robots can play important roles in construction of the human social life. The closer robots come to human beings, the more likely it is that robotic ethical risks will emerge.

During the twentieth century, various negative ethical effects have been linked to advanced modern technologies such as nuclear physics, bioengineering and computer science. For example, the development of the computer industry, as one of the most prosperous industrial ventures, simultaneously brought with it a great deal of unintended ethical challenges: job displacement, privacy concerns, intellectual property disputes, alienation from the real world, security fears, cyberbullying, telecommunication fraud, internet addiction and so on.[5] The governments of many states were reproached and blamed for these negative effects by their publics. These states then either closed down or imposed restraints on the scientific research and technological applications in question in order to alleviate public concerns and fears.[6] The lessons learned from the history of computers requires us to take seriously the ethical and social challenges that will arise from robotics in order to sustain

他风险隐而未显，但并非不可预见。为了叙述方便，我们将根据机器人的不同用途来展示它们引发的部分伦理后果。在机器人技术应用带来的众多社会伦理问题中，其中一些具有特殊性，另外一些则是共同的。

工业机器人会在全球经济竞争中增加产业盈利并提升竞争力。然而，历史上不乏工业机器人造成的严重安全事故。1979年，美国密歇根州的福特制造厂内，机器人在整理仓库时因手臂撞击一名工人头部而致其死亡。[7] 1981年，日本川崎重工制造厂一名工人进行维护作业时，由于匆忙而未有效关闭机器人，被机器人杀死。[8] 其次，机器人擅长沉闷、肮脏、危险的工作，其效率往往优于工人。然而，工厂大量使用机器人会削减现有工作岗位数量并造成工人失业的焦虑和恐慌。最后，机器人能够极大提高生产线效率。但是，如果机器人主导生产流程，这不仅会造成生产环境非人化，而且会导致生产人员自尊心受挫。即使在今天盛行的协作机器人那里，生产环境中人机交互的增加也伴有人与人之间交流减少的隐患。[9]

医用机器人在手术、康复和护理领域的应用正在迅速增长。现在世界上使用最广泛的远程控制手术机器人 Da Vinci 机器人系统能够极大减少手术创伤、提升手术精准度并缩短术后康复周期。然而，首先，手术机器人本身就存在着一定技术局限和技术风险。其次，机器人手术一旦失败会引发医疗责任归属问题。再次，手术机器人会带来更加复杂的患者个人数据隐私监管隐患。最后，现有手术机器人的高额成本导致机器人手术无法惠及全民并造成医疗资源分配

a beneficial programme of robotics. Because of the rapid growth of robotics and close human-robot interaction, the ethical challenges that emerge from robotics are far more difficult and threatening than nuclear physics or computer science. In fact, some unintended negative accidences have already occurred while others are still imminent. While some of the ethical problems seem to be common to all types of robots, for the purpose of our concise presentation here, we will introduce and classify some of the ethical issues that emerge from the diversity of robot deployments.

Industrial robots are likely to increase economic profits and intensify manufacturing competition in the global marketplace. However industrial robots have also been responsible for severe safety breaches. In 1979, in a Ford factory in Michigan, a robot's arm that was retrieving parts of warehouse hit the head of a worker and killed him.[7] In 1981, a factory worker was killed when carrying out a maintenance work on a robot at a Kawasaki plant.[8] Again, robots are good at 3D works (that is, work that is dull, dirty and dangerous) with more efficiency than factory workers. Yet the untimely introduction of numerous robots may decrease the number of current employment opportunities and cause worry and fear over unemployment. Finally, robots can increase production at a plant. But robotic dominance on the production line will dehumanize the working environment and do injury to workers' self-esteem. With robots becoming increasingly in vogue, the increase of human-robot interaction may reduce opportunities for human-to-human communication.[9]

The demand of **medical robots** is rapidly growing in the areas of surgery, rehabilitation and healthcare. The da Vinci robotic surgical system as one of the most popular tele-robots in the world can minimize surgical invasion, improve surgical precision and shorten the process of recovery. First, like any other technological artefacts, surgical robotic systems cannot avoid certain technological limits and risks. Secondly, in the event of surgical failure, the complicated issue of accountability may emerge. Thirdly, surgical robotic systems may increase the risk of private data breaches for patients. Finally,

不均。另外，手术机器人的巨额商业利润可能会间接影响患者手术知情同意权。[10] 在一些国家的医院中，护理机器人已经开始获得应用。然而，护理机器人既无法恰当回应病人的情绪也无法在不同呼叫中分配回应优先次序。此外，护理机器人有可能带来患者对机器人的情感依赖。[11]

家用机器人在全球机器人市场上的产品种类仍然比较有限，主要提供清扫地面、除草、教育和娱乐等简单功能。今天，人们对于不同功能的家用机器人需求与日俱增。相信在不久的将来，会有功能更强大的机器人进入更多家庭，成为人类家居用品或家庭伴侣。然而，机器人与人的亲密相处首先会进一步加剧个人隐私泄露的风险。其次，机器人滥用的风险也在成倍增加。再次，密切的人机交互会造成人对机器人单向情感依附并带来被操纵利用的风险。最后，随着智能化和自主性功能的增强，家用机器人自身的权利问题也会逐渐成为新的伦理关切。

特种用途机器人用于特殊环境，执行专门任务，包括无人驾驶汽车（轮式移动机器人）、救援机器人、深海机器人、探月机器人、军事机器人等等。近年来，无人驾驶汽车的发展引发了一系列严重的安全问题。2016年，一位美国男子在驾驶特斯拉无人驾驶汽车(Tesla Model S sedan)时没有按照合同要求对自动巡航系统进行监控，与前方卡车惨烈相撞而身故。[12] 2010年，美国军方无人机试飞过程中失控，侵入华盛顿特区禁飞区，对白宫和其他政府部门构成严重威胁。[13] 除了安全问题，无人驾驶汽车和军事机器人还会引发商业

since the high costs of robotic surgery are not affordable by ordinary people, they result in an unequal distribution of medical resources. In addition, the enormous profits gleaned from surgical robotic systems may indirectly have negative consequences for the human right of informed consent.[10] In some countries, nursing robots have been introduced. But nursing robots can neither respond appropriately to patients' emotional needs nor prioritize among demands made by patients at the same time. It is also possible that the use of nursing robots may result in a one-sided emotional bond.[11]

Domestic robots are still today limited to a few types and functions. They are deployed to attend to rather simple functions in education and entertainment, and as floor cleaners and mowers. Yet the demand for diverse functions of domestic robots is increasing. It is certain that in the near future more powerful robots will enter the home as household facilitators or family companions. But intimate living arrangements of robots with humans may increase the risk of private data breach. Secondly, the risks of misusing robots may increase many fold. Thirdly, the one-sided emotional bond that results from the intimate interaction of humans and robots make people more vulnerable to illegal exploitation. Finally, the increase of robotic intelligence and autonomy may give rise to a new ethical concern with respect to the rights of the domestic robot itself.

Robots for special use (such as self-driving cars, rescue robots, outdoor robots, military robots and so on) can perform particular activities in special environments. In recent years, self-driving cars have already been involved in several traffic accidences. In 2016, an American man driving a Tesla Model S sedan who did not override the autopilot system died in a violent crash with a truck.[12] In 2010, an American army drone lost control, entered the Washington no-fly zone, and threatened the White House and other departments of state.[13] In addition to safety problems, self-driving cars and military robots may give rise to serious challenges in business and war ethics. Again, outdoor robots can enter areas on the planet that are inaccessible to human beings. Although they

伦理以及军事伦理方面的挑战。此外，特种用途机器人能够探测地球上无人进入的领域，接触无人接触的生物。机器人的这类户外探测活动在带来巨大经济利益的同时也会加剧环境伦理方面的忧虑。

机器人业的科技研发正在日新月异。机器人引发的伦理挑战将广泛涉及工业、军事、法律、宗教、医疗、保健、心理、性、隐私等各种不同社会领域。大量意想不到的社会伦理问题正在随之涌现，彼此交叠，错综复杂，向人类提出严峻的挑战。面对机器人这个新兴科技产业的蓬勃发展，国际社会相关伦理政策的制定迫在眉睫。

1.3 BSI 和 IEEE 机器人伦理的经验与局限

世界机器人业正在成为深刻变革人类生产和生活的关键力量。机器人业的主要参与者都开始采取主动措施，积极研究和制定普遍有效的机器人伦理政策。2016 年，联合国大会通过声明，要求应该禁止研发"缺乏人类控制的自主武器系统"。联合国还在海牙特别成立了专门研究机构来研究机器人学和人工智能的治理。在国家层面，英国 BSI 于 2016 年发布《机器人与机器人设备》作为机器人和机器人系统设计和应用的伦理指南。2015 年，欧盟议会建议欧盟委员会制定机器人学的规范。2017 年欧盟议会通过的《机器人民法条例》建议报告中同时包含《机器人学宪章》《机器人学工程师伦理行为准则》以及《研究伦理委员会准则》。在国际专业机构层面，IEEE 分别于 2016 年和 2017 年公布了两版机器人伦理政策性文件《与伦理协调的设计》，并计划在广泛争取国际意见的基础上于 2019 年发布最终版。

may produce a huge profits, the robotic exploration of inaccessible territories cannot but cause environmental worries over the intrinsic values of nature.

The robotic technologies and industry are developing rapidly. Ethical challenges from robotics are of concern over the broad range of human activities such as manufacturing, military, law, religion, medical treatment, healthcare, psychology, sex and privacy. An enormous number of ethical problems are on the horizon. Ethical problems from robotics are so intertwined and complicated that human society is facing unprecedented challenges. In response to the exponential growth in the emerging robotic technologies, the international society is urged to investigate this phenomenon thoroughly and elaborate an implementable ethical strategy.

1.3 Lessons Learnt from BSI and IEEE

The global robotics industry is becoming a powerful force in transforming all aspects of human productivity and life. Leading participants in the field have initiated forums to actively discuss robot ethics and formulate policies that have universal validity. In 2016, the United Nations General Assembly made a statement requiring the prohibition of "autonomous weapon systems that do not need significant human control". In addition, the UN made a special effort to establish a technical institution in the Hague for inquiring after the governance of robotics and AI. At the national level, the British Standard Institution (BSI) released *Robots and Robotic Devices* in 2016 as ethical guidelines for the design and application of robots and robot systems. In 2015, the EU Parliament suggested that the EU Commission enact normative standards for robotics. In the draft report of *the Civil Law of Robots* submitted by EU Commission in 2017, three other documents were also appended, namely, the *Charter of Robotics*, the *Ethical Code of Conduct for Robotics Engineers*, and the *Ethical Committee Guidelines for Research*. At the level of international professional institutions, IEEE published two versions of its policy document, *Ethically Aligned Design*, in 2016 and 2017 respectively, and following the further consideration and adoption of relevant opinions solicited from around the

2018年欧盟委员会人工智能高层专家组公布了《可信赖人工智能伦理导则》初稿，开始广泛征集利益相关者的公众意见。这些不同的国际努力既为今天中国机器人伦理标准化工作提供了宝贵经验，也在一定程度上显露了既有伦理框架在面对机器人这个新兴科技时的局限。

在正式发布的《机器人与机器人设备》中，BSI的机器人伦理标准主要从工程伦理角度，利用英国在工业／工程设计方面的现有标准（主要涉及安全性和可靠性）来考虑机器人设计所要满足的一般伦理要求。作为工程设计的一个特殊领域，机器人的设计和研发不仅在性能上要安全可靠，而且要充分尊重人类和自然价值。BSI的机器人伦理标准从社会、商业／金融、机器人使用／滥用以及环境这四个方面来考察机器人可能带来的伦理风险。BSI建议在用户导向设计和用户体验确认有效之外补充机器人设计者的伦理教育，并通过恰当选择机器设计指标来规避伦理风险和减少伦理伤害。BSI伦理标准着重关注机器人设计为个人、社会和环境带来的负面安全性问题，较少涉及机器人对于人类既有伦理秩序的正向调整影响。虽然BSI伦理标准提到了伦理原则冲突的可能，但是它还没有提出可供讨论的解决策略。另外，BSI伦理标准只是对于迄今已出现和预期的机器人伦理问题的初步归类，目前仍然缺乏相应的系统性伦理框架。

world, projects that it will publish its final version in 2019. In 2018, the European Commission's High-Level Expert Group on Artificial Intelligence announced the *Draft Ethics Guidelines for Trustworthy AI* for public feedback from all the stakeholders. Today these endeavours made at a national and international level, provide valuable experience that can inform the formulation of the Chinese ethical code for robots, while they also, to some extent, show the limits of the existing ethical frameworks in the face of the emerging robot technology.

In its officially released *Robots and Robotic Devices*, BSI reflects on general ethical standards that the ethical design of robots needs to satisfy, approaching the issue largely from the perspective of engineering ethics, and adopting existing British standards in industrial/engineering design (that are mainly concerned with safety and reliability of products). BSI is explicitly aware that the design and development of robots, as a special field of engineering design, needs not merely to ensure safety and reliability in the performance of products, but also needs to pay adequate concern for human beings and for the value of nature. The BSI's standards for robot ethics reflects upon ethical risks that the design and application of robots would probably occasion from four perspectives: societal, application/misuse, commercial/financial, and environmental. BSI suggests that the ethical education of designers be supplemented in the design of robots in addition to the need to take into account user-oriented designs and product validity confirmed by users' experience. It also suggests that ethical risks be avoided and ethical harms be reduced by properly choosing the indexes for machine design. BSI's ethical standards pay attention more to the negative problems of safety that the design of robots occasions, and less to the effects of the positive adjustment of robots to the existing human ethical order. Even though BSI's standards do mention the possibility of conflict between ethical principles, they fail to offer strategies for solving conflicts that might be further discussed. In addition, BSI's ethical standards are only a preliminary classification of the ethical problems that the design and development of robots have given rise to thus far and that can be anticipated at this juncture, and thus falls short of an applicable and systematic ethical framework.

IEEE伦理政策从人类中心立场出发，更加细致地分析了机器人技术对人类生活和人类价值可能产生的影响。IEEE所持有的人类中心立场是指机器人以及相关技术是为人类利益而发展出来，机器人的研发和设计应该服务于人类广泛分享的伦理理想并促进全人类福祉。IEEE的伦理政策把人类能够分享的核心价值作为设计和研发机器人的根本条件并提出了五个一般原则：人权、福祉、追责、透明度和滥用意识。就机器人的设计而言，IEEE采纳了"在伦理约束下优化效益"的指导方针，在通过伦理约束来确保机器人不会对人类造成伤害的前提下，优化机器人对促进人类福祉方面产生的益处。IEEE政策文件《与伦理协调的设计》前后两版中的变化正确凸显了对于人类文化多元性和人类价值多元性的关注。IEEE的伦理政策同时倡导一种促进人类福祉的整体论设想。目前，IEEE的机器人伦理政策草案仍然缺乏一个整合性框架来同时兼容人类福祉与其他人类价值和自然价值。另外，IEEE的人类中心框架似乎很难有效地应对具有增强人类能力的机器人技术（例如赛博格）带来的伦理挑战。

近年来，机器学习和机电一体化技术突飞猛进，机器人智能和自主能力极大增强，人机交互愈加密切。机器人在增强人类能力方面的应用、机器人自身的伦理身份、机器人与人的伦理关系以及机器人与自然环境之间的关系都不断呈现出前所未有的复杂特征。全球机器人业的理论突破和技术创新要求我们必须审慎对待其他国家现有机器人伦理框架和标准的局限。中国文化传统拥有丰富的伦理

Taking a human-centric standpoint, IEEE's ethical policy is based upon a careful investigation of the potential impacts that robotic technology will have on human life and human values. The human-centric standpoint taken by the IEEE means that robots and related technologies are to be developed with the purpose of serving human interests, where the design and development of robots should confirm to the ethical ideas widely shared by human beings and accordingly should promote the well-being of all mankind. IEEE's ethical policy takes the core values that human beings share to be the essential conditions for designing and developing robots, and goes on to propose five general principles: human rights, well-being, accountability, transparency, and awareness of misuse. With respect to the ethical design of robots, IEEE adopts the guideline of optimizing benefits under ethical constraints, namely, maximizing the benefits robots can produce in promoting human well-being under the condition that the built-in ethical constraints will ensure that robots will not do harm to human beings. From some changes introduced as the first version of IEEE's *Ethically Aligned Design* evolved to become its second version, it can be seen that IEEE wants to highlight its attention to the plurality of human cultures and human values. Meanwhile, IEEE's ethical policy also advocates the promotion of human well-being in a holistic manner. At present, an integrated framework capable of accommodating human well-being within other human values and the value of nature is still lacking in IEEE's policy for robot ethics. In addition, it would seem that IEEE's human-centric framework is inadequate to face effectively the ethical challenges arising from robotic technologies that will enhance human capabilities (bio-chemistry cyborgs, for example) brings about.

In recent years, as technologies of machine learning and advanced mechatronics have been advancing rapidly, the intelligence and autonomous capacity of robots has been vastly enhanced, and the interaction between man and computer has also become more and more intimate. Breakthroughs and innovations in robotics and related sciences and technologies make it necessary that we deal deliberately with the limitations of existing frameworks for robot

思想资源。我们希望借助中国伦理传统制定一套具有广泛包容性的机器人伦理体系，为世界机器人设计、生产和监管贡献可被共同使用的国际性伦理框架。

ethics. The Chinese cultural tradition is rich in its ethical resources. It is our goal, with the resources that traditional Chinese ethics provides, to establish a highly comprehensive framework for robot ethics, and to thus make a contribution to the development of an international ethical framework that can be commonly used in the design, production and governance of robots all over the world.

第二章 中国机器人伦理宗旨
Chapter 2　The Ethical Commitment of Chinese Robotics

无论在生产还是生活领域，机器人正在以我们无法想象的方式彻底变革人类社会。世界各国机器人业的发展都拥有自己独特的法律、政策、伦理和经济环境。今天，在中国和其他很多国家地区，人们日益达成共识，认为"财政成功指数"无法全面而充分地衡量人类社会的繁荣程度。除了利润、GDP、消费指数、就业以及经济增长这些经济指标之外，中国机器人业同样要致力于具有伦理驱动和社会责任担当的机器人系统设计和研发。在即将到来的机器人革命时代，中国机器人业必须认真对待机器人相关的复杂伦理挑战和巨大社会影响。在此，中国机器人业郑重宣布：在为世界机器人科技革命做出自己贡献的同时，我们将始终对自然秩序的和谐和人类生活的美好价值承担起足够的伦理关怀和社会责任。

由于全球盛行的自动化潮流以及工业机器人领域持续的技术革新，中国的工业机器人需求显著提速。当今，制造和生产的本质正在发生深刻改变。简约化、数据化以及协作化已经被视为未来工业机器人发展所必须关注的三个核心技术要素。[14] 国际机器人联合会（IFR）2017年的统计数据显示[15]，在 2011—2016 年间，全球工业机器人年均销售量约 21.2 万台，年均增长率 12%。在全球各个地区中，亚洲工业机器人增长势头最为强劲；中国、韩国和日本位居 2016 年

The technological advance in robotics is so thoroughly transforming forms of manufacturing and social life that we could never imagine before. Robotics often develops against a backdrop of law, policy, ethics and economics among other social, cultural and political forces. Due to a growing consensus on the insufficiency of the "fiscal metrics of success" for the measurement of the prosperity of human society both in China and other countries, apart from profit, gross domestic product (GDP), consumption levels, occupational safety and economic growth, Chinese robotics is committed to a socially responsible and ethically driven design and development of robot systems. Both ethical considerations and the social impact must be seriously taken into account in the coming robot revolution. Chinese robotics is firmly committed to make its technological contributions to the coming robot revolution in a way full cognizant of its social responsibilities and with the ethical care needed to guarantee an emerging harmonious natural order and the value of human life.

The Chinese demand for industrial robots has accelerated due to the ongoing trend toward automation and the continued technological improvement in industrial robots. Today the nature of the workplace and manufacturing is changing. Simplification, digitalization and collaboration are regarded as three core factors in the development of industrial robots.[14] IFR reported in 2017 that between 2011 and 2016 the average annual supply of industrial robots rose to about 212,000 units with the average sales increase as 12% per year worldwide.[15] Among other regions, Asia is the world's strongest growth market.

全球工业机器人销量前三位。从 2013 年起中国就已经跃居成为全球最大的工业机器人市场。2016 年，中国工业机器人销售已达 8.7 万台，接近欧美工业机器人销量总和，占当年全球工业机器人销售市场总份额的 30%。根据国际机器人联合会预测，从 2017 年到 2020 年，全球工业机器人年均增长率将至少保持在 18%，总量 1700 万台工业机器人将装配在全球众多工厂。除了传统工业机器人带来的安全与保护问题，工业领域日益增加的人机协作系统引发了众多新的伦理关切，例如劳动人口失业、工作区域和流程非人化以及生产人员交流减少等。

今天机器人已不再仅仅服务于工厂环境，而开始广泛进入千家万户的日常生活。虽然服务机器人仍处于蓬勃发展的初期，但是它们的尝试应用已经在众多行业领域造成了深远影响。目前，专业服务机器人的增长主要集中在物流、国防、作业以及医疗行业；家用机器人的增长则集中在地面清洁、除草、教育娱乐以及残障人士助力领域。国际机器人联合会 2017 年的统计数据显示[16]，专业服务机器人生产主要集中在美国和欧洲。与前一年度相比，2016 年亚洲专业服务机器人销售增长达 72%。根据国际机器人联合会的统计，2016 年，亚洲和澳大利亚教育机器人占全球市场供给份额的 60%，老年和残障人士助力系统则占到 94%。随着人口老龄化问题的逐步加剧，全球对于个人或家用机器人的关注和需求将与日俱增。在中国，人们对于物流系统机器人、作业机器人、公共关系型机器人、人体能力改善或增强的外骨骼机器人、医用机器人以及国防机器人

In 2016, China, the Republic of Korea and Japan were the top three robot markets in the world. Since 2013, China has already risen to the world's leading position in the marketing of industrial robots. In 2016, about 87,000 industrial robots were sold in China. According to IFR statistics, this sales figure comes close to the total sale volume of Europe and the Americas combined (97,300 units). Robot installations in China occupied a 30% share of the total supply in 2016. According to IFR estimates, from 2017 to 2020 global robot demand is projected to increase by at least 18% with a total of 17 million industrial robots being assembled in factories around the world. Apart from the traditional issue of safety and security, the increased collaboration of humans and robots in the workplace creates other ethical concerns such as unemployment, dehumanization and reduced human communication.

Today robots have moved beyond the floor of workforce and entered into our daily life. Despite the fact that the development has only recently begun, professional service robots have already made significant impacts in areas such as agriculture, surgery, logistics and underwater application. The demand for professional service robots is growing in the areas of logistics, defence and medicine. Personal and domestic robots have also experienced a global increase with the production of a few products: floor-cleaning robots, robo-mowers and robots for "edu-tainment". IFR's incomplete statistics[16] show that the major portion of all units of professional service robots was produced in America and Europe. Compared with 2015, the sales of professional service robots in Asia increased by 72%. In the same year, 60% of all reported entertainment robots and 94% of elderly and handicap assistance systems were produced by Asian and Australian companies. Because at the global challenge at an aging population and other critical factors, service robots are in the near future expected to have increasingly broad utility in a large range at areas. In China, there is an increasing

的需求越来越大。机器人将摆脱传统的3D类型（枯燥、肮脏和危险）任务来进入人类寓居空间并和我们产生更加直接的合作交流。一旦机器人进入人类环境并获得和人类更加密切的合作，个人和社会生活图景将受到前所未有的冲击。难以估量的伦理和法律问题也将接踵而至。

为了促进机器人业的健康发展，中国与世界其他国家持有很多相似的立场。中国机器人业相关者同样充分地认识到，如果我们不能有效处理机器人引发的众多伦理后果，中国乃至世界机器人业都将无法真正获得繁荣发展。

对于像机器人这样的新兴科技，仅仅遵循法律和政策是根本不够的。一个简单的原因在于相关的法律和政策仍然付之阙如。另外，法律和伦理的范围也并不总是彼此重合。历史经验显示，人类社会从不欢迎危害自然和人类自身的科学技术运用。机器人的设计和应用亟须引入恰当的伦理标准，这已经成为全球机器人业的基本共识。机器人设计和应用的伦理标准不应只是限制性和否定性条例，而应成为能够融入和驱动机器人技术发展中的前瞻性导则。《中国机器人伦理标准化前瞻（2019）》的制定和颁布正是中国机器人业积极行动的先导，表达中国政府和中国公众对于机器人业的深刻伦理关切，推动并培育全社会对于机器人的理性认知和健康讨论，为未来机器人业伦理标准、指南或细则制定开辟道路，进而推动中国机器人业产学研结构的升级调整。

在准备这部《前瞻》的过程中，我们全面而广泛地参考了其他

demand for logistic systems, defence robots, field robots, public relations robots, powered human exoskeletons as well as medical and care robots. Whether in industry or in the home, the trend is for increasingly intimate collaboration between human beings and robots. Robots are going to change human lives to an extent never before imaginable. The more robots enter the human environment, the more complex the ethical challenges will become.

With respect to the healthy development of robotics, Chinese shares many concerns with other countries. The Chinese stakeholders along with the world of industry recognizes the fact that unless we address the numerous ethical concerns and problems that robots give rise to, neither China nor the world will be able to develop a flourishing robotics industry.

For an emerging technology like robotics, merely following law and policy cannot be enough for no other reason than they are not yet available. In addition, the substance of law and ethics may not always overlap with each other. In the long run, the development of harmful technologies is never welcomed by human society. It has become a consensus among international stakeholders in the robotics industry that an ethical dimension must be embedded in the design and deployment of robots. Ethical rules governing future robotic design and deployment should not be merely veto and prohibitive, but rather in a prospective way embedded in and even drive the technological advance in robotics. This document is meant to be an initiative that declares the deep concern on the part of the Chinese government and Chinese society, to promote and cultivate a rational discussion in the entire country, and to pave the way for establishing future ethical standards, guidelines and regulations as well as to strengthen the foundation of the Chinese robotic enterprise.

When preparing the *Prospects*, we made a broad reference to the existing documents on robot ethics in other countries. It turns out that as with other

国家和地区已经制定和颁布的机器人伦理文件。综观这些文件，所有机器人的设计既要依赖它们所履行的核心任务，也要像其他人造制品一样满足安全可靠的标准。机器人必须具有可以为用户在其所必需程度上能够明确理解的使用规则或操作指令。

然而，机器人与传统机器以及其他类型的人工智能产品也存在鲜明的差别。机器人的独特之处在于，不仅具备一定程度的智能和自主能力，而且具有物理外形和移动能力。因此，机器人的运行本身就会造成特殊的伦理后果，带来其他工程和工业设计领域不曾遇到的伦理难题。一般来说，机器人的自主性越强，可能招致的伦理风险就越大，为了应对风险而需要考虑的伦理条件就越复杂。今天，国际机器人业的相关者日益清楚地意识到，人们要在对机器人自主性的追求和其可能导致的各种伦理风险之间取得平衡。

另外，机器人的设计和研发不仅要考虑它们所能带来的经济效益，也要考虑它们对人类生活整体和自然环境可能造成的不可逆的影响。我们不仅希望机器人的发展和利用不会危及人类和自然，而且希望它们能够有效地促进人类福祉并维护良好的自然生态。

目前，国际机器人业相关者已经提出的伦理标准，或者基于以个体自主为根基的近代伦理体系（义务论或后果论），或者依赖受古希腊亚里士多德传统启发的美德伦理学体系。即使一些系统化的机器人伦理标准制定工作尝试理解世界多元的文化传统，非西方的伦理传统也往往囿于西方传统自身的问题视角而无法真正贡献可被普遍认可的创造性伦理框架。鉴于既有机器人伦理框架普遍存

artefacts, the design of robots must not only be oriented toward targeted tasks but also accord with existing safety requirements. For any type of robot, it is imperative that guidelines for use and a code of operation be accessible and made clear to user in a required way.

Yet robots are remarkably different from traditional machines and other types of non-embodied artificial intelligence. Robots are special in that they not only incorporate intelligence and autonomy but are also embodied and capable of movement. Robotic operations are thus likely to have ethical consequences. Some of the effects may involve ethical dilemmas that have never arisen in other areas of engineering and industrial design. Generally speaking, the degree of ethical risks that arises from the use of robots is proportionate to the increase of robotic autonomy. It follows that an increasing range of ethical considerations must be taken into account to mitigate the ethical risks. Today stakeholders are increasingly aware of the fact that we need to maintain a balance between the various ethical risks and our technological pursuit of robotic autonomy.

Moreover the design and deployment of robots should not be determined merely by economic and financial profits, but also by the irreversible effects they will have upon the unity of human life and the natural environment. It is certainly our hope that the development of robotics will not harm and endanger the world. Indeed we fully expect that robotics can be deployed efficiently to promote the well-being of human life, and to maintain nature's beautiful ecological system.

The existing international documents and standards of robot ethics are grounded on either the modern ethical principle of individualism (deontology or consequentialism) or neo-Aristotelian virtue ethics. Even though an international institute such as IEEE has become increasingly aware of the

在的人类中心主义理论局限，《中国机器人伦理标准化前瞻（2019）》尝试从中国伦理传统中寻找独特启示，并在充分包容人类现代价值和其他伦理观念基础上发展一套可资借鉴和使用的国际性机器人伦理标准化体系。

为此，中国机器人伦理标准化体系要发展一套**去人类中心主义**的伦理系统，坚持良性共生秩序及其所要求的整体论含义的善。中国机器人的设计和发展要同时致力于促进自然价值和人类价值。**中国机器人伦理标准化体系的核心目标，就是要把对于多元价值的尊重和对于共生结构的优化充分融入在未来机器人技术和产品的研发与创新中。**

plurality of ethical cultures and traditions, the distinctive perspectives of non-western traditions are often limited by western ways of analysis and hence are difficult to grasp fully. To the extent that the ethical limitations in existing documents may be threatening to the future of robotic development, the *Prospects* search for unique inspiration from the Chinese classical traditions. By taking full measure of and integrating modern human values and the ideals of other cultures, the Chinese system for the standardization of robot ethics as developed here seeks to be an international resource and instrument that can be borrowed and applied in other communities at the same time.

Uniquely inspired by the classical Chinese and many other valuable cultural traditions, the Chinese system for the standardization of robot ethics rejects the anthropocentric standpoint as widely adopted in the other robot ethical documents. Chinese robotics insists on the holistic good of a well-ordered world. Chinese robotic design and deployment is thus committed to promote both natural and human values at the same time. It is the single goal of Chinese robotics in its technological innovations and products to embody a respect for a plurality of values and the optimizing of a symbiotic structure.

第三章　中国机器人伦理标准化体系

Chapter 3　The Chinese System for the Standardization of Robot Ethics

《中国机器人伦理标准化前瞻（2019）》旨在为未来中国机器人伦理标准制定提供一套具有科学性、普遍性和可操作性的标准化体系。

在中国和很多国家地区，古典和现代传统成为人们生活中并存的思想资源。它们既有不少共同之处也有相当多的分歧。这部《前瞻》借助中国古典伦理传统中的理性资源，来建立一个整合性的伦理标准化体系。我们这里所提出的中国机器人伦理标准化体系并不是一套在具体案例中直接应用的逻辑推导原则，而是在未来中国机器人业伦理指南或细则制定中所必须遵循的善的整体框架。由于中国多元的古典传统能够在其他许多文化共同体中找到重叠共识，这部《前瞻》所提出的整体框架同样可以作为"不同社会团体和信仰所共同分享的科学、文化和技术工具"。[17]

准备这部《前瞻》的过程中，我们已经注意到国际上开始的广泛争论，机器人的决策是否能够具有道德价值以及机器人是否是像人一样的道德智能体。与此相关，机器人的伦理既可以指关于机器人研究者、生产者和用户的伦理，也可以指写进机器人程序中的道德编码，还可以指关于具备道德推理和自我决策能力的机器人本身的伦理。[18]由意大利机器人学家韦卢奇奥（Gianmarco Veruggio）最早提出的"机器人伦理"（roboethics）这个术语就是指机器人设计者、

The *Prospects* are meant to develop a scientific, generalizable and practical system for the standardization of ethical rules governing Chinese robotics in future.

In China and elsewhere, both classical and modern ethical traditions co-exist as living resources for the lives of its people. This document borrows from Chinese classical traditions to develop an integrative ethical system. The programme proposed here does not aim at establishing a set of ethical principles to be universally applicable in all particular cases. Rather it proposes to provide an overall framework of the good as it is implicated in the ethical guidelines, standards and regulations of robotics. Insofar as there is a great deal of consensus between Chinese classical traditions and the values of many other cultures, the proposed overall framework can also be referred to as "scientific/cultural/technical tools that can be shared by different social groups and beliefs".[17]

This document is informed by a number of debates on the possibility of attributing moral values to the decisions of robots, and the characterizing of robots as moral agents like human beings. Robot ethics can range from an ethics for the robotics researchers, the producers and the users of robots, to the moral codes programmed into robots, and to that of robots capable of ethical reasoning and self-decision.[18] The term "roboethics" coined by the Italian roboticist Gianmarco Veruggio refers to the ethics of the designers, manufacturers and end users of robots. The foremost concern of the document is on roboethics for the existing technology, although it seeks to be capacious

制造者和终端用户的伦理。鉴于现有的机器人技术条件，这部《前瞻》首要关注的就是机器人伦理。但这并不意味这里提出的伦理框架仅仅局限于机器人设计者、制造者和终端用户伦理。事实上，这部《前瞻》所提出的伦理框架也为尚不可预期的人工道德智能体的自主能力增强预留了足够的伦理讨论空间。[19]

在这一章中，我们将首次提出并详细解释中国机器人伦理方案。我们既要说明这个方案的中国传统伦理根据也要展示它在全球各个国家的普遍可适用性。中国机器人伦理方案一方面在整体性善的基础上提出具有普遍性的元伦理框架，另一方面在今天国际共识基础上把这个元框架具体化为五项伦理标准。中国机器人伦理方案并非空洞的宣言，而是具有良好的可操作性。为此，我们设计了五步骤程序来支持和确保这个伦理方案在未来中国机器人设计和研发中得以执行。

3.1 中国优化共生框架

在世界各国，机器人业都已经被视为变革生产制造和人类生活的关键性转型技术。根据国际标准化组织（ISO）的最新界定，机器人被定义为"具有一定程度的自主能力，可在其环境内运动以执行预期任务的可编程执行装置"[20]。机器人不仅是一套智能系统，而且必须具有物理外形和运动性。根据用途，ISO把机器人划分为两种类型，工业机器人和服务机器人。经过半个多世纪的发展，工业机器人对于经济增长、创造就业岗位和确保产业竞争力具有举足轻重

enough to cover future possible and unanticipated developments with respect to artificial moral agents (AMAs).[19]

In this section, we will introduce and explicate the Chinese Optimizing Symbiosis Design Programme. We will trace the Chinese roots and affirm the universal applicability of the programme. In addition, we will also work out a five-step process to implement the programme in future robotic design and development.

3.1 The Fundamental Framework of the Chinese Optimizing Symbiosis Design Programme (COSDP)

Today robotics is regarded as a key transformative technology that can revolutionize manufacturing and transform the daily lives of people. A robot is defined as an "actuated mechanism programmable in two or more axes with a degree of autonomy, moving within its environment, to perform intended tasks."[20] Robots are divided into two types, namely, the industrial robot and the service robot. After more than a half century of rapid development, the industrial robot has become essential to the future of economic growth, the creation of new jobs, and the need to ensure competitiveness in the market place. In the past decade or so, diverse forms of professional and consumer service robots have become more available to us. Because of the global challenge of an aging population and other critical factors, service robots are expected in the near future to have increasingly broad utility in a large range of areas such as logistics, transportation, surgery, rehabilitation, healthcare, education, household assistance, entertainment, military service and so on. Robots will develop far beyond the traditional 3D tasks (dirty, dull and dangerous) to perform assignments currently occupied by human

的作用。近十来年，不同类型的生产和服务机器人获得越来越多的应用。由于人口老龄化和其他重要因素带来的全球性问题，服务类机器人将在物流、运输、手术、康复、医护、教育、家政、娱乐、军事等领域逐渐获得广泛应用。机器人将走出传统的"3D"岗位，通过与人们的协作来完成目前由人自己执行的任务。随着机器人技术迅猛发展和广泛使用，难以估量的伦理和法律问题将接踵而至。全球机器人业亟须明确的机器人伦理策略来应对即将到来的风险和挑战。

今天，机器人革命的伦理挑战尚未完全到来。然而，基于大众影视和科幻文学来玄想机器人带来的伦理困难和应对策略即便不危险也是无用的。为了规避科幻玄想对机器人业发展的潜在阻碍，这部《前瞻》的制定始终瞄准既有的科技和在不远的将来可预见的发展突破。作为中国机器人伦理标准化体系，这份文件包含足够的理论弹性来面对未来难以预料的技术创新。迄今为止，古希腊的亚里士多德主义、印度的佛教、非洲的乌班图（Ubuntu）、日本的神道教以及现代世界的义务论和后果主义伦理传统都被相继援引，作为世界各国应对机器人伦理挑战的重要思想资源。中国的多元文化传统具有异常丰富的伦理思考和理论体系。为了让中国思想传统向中国乃至世界机器人业的长足发展贡献自己的伦理洞见，我们在此提出**中国优化共生设计方案**（Chinese Optimizing Symbiosis Design Programme 或者 COSDP）。

中国优化共生设计方案是植根于哲学伦理的机器人伦理标准化体系。它既实质性地汲取中国传统伦理的思想资源，又广泛吸纳在

beings and often in direct collaboration with us. As robots continue to enter the human environment and to find closer interaction with humans, they will have an unprecedented impact on individual and social lives. This changing environment will give rise to many ethical and legal problems, and to many unforeseen challenges.

Since such full-fledged ethical challenges in the robot revolution have not yet arisen, it is useless if not dangerous to speculate on ethical solutions on the basis of popular movies and literature. This document is composed with an eye to the existing technology and to its foreseeable advances in the near future. As a general system for the standardization of robot ethics, the document seeks to contain enough flexibility to be able to deal with unexpected innovations in robotics as they occur. So far, Aristotelianism, Buddhism, African Ubuntu tradition, Japanese Shintoism and modern deonology as well as consequentialism all have been referred to for our confronting ethical challenges from the development of robotics. Diverse traditions in China may equally provide us with rich sources of ethical thoughts and theories. For the socially responsible and ethically driven design of robot systems, we propose the Chinese Optimising Symbiosis Design Programme (COSDP).

The Chinese Optimizing Symbiosis Design Programme is rooted in a philosophical and ethical framework that draws substantially on Chinese classical traditions while at the same time taking into account a wide swath of contemporary theories, debates and practices surrounding the technologies of robotics as they have emerged in other countries and cultures. The term "optimizing symbiosis" reflects certain persistent traditional Chinese cultural assumptions and values: the holistic, ecological nature of the human experience, the high value of integration and inclusiveness, the *yinyang* interdependence of

其他国家中发展出的当代机器人伦理理论、争论和实践。我们提出的"优化共生"概念准确地反映了中国文化传统中一些稳定和持存的价值与信念：整合与包容的最优先性，人类经验的整体性和生态性特征，在环境中所有事物的阴阳相生，通过共享多元渴求在和谐中最大可能地保留差异，以及既非僵化也非目的论式的生生不息的自然、社会和政治秩序。中国优化共生设计方案希望成为中国以及其他各国机器人伦理标准化的可能参考。

中国优化共生设计方案包含如下三个从低到高的理论层级：整体论含义的善，四个元向度以及五个主要伦理目标。其中整体论含义的善和四个元向度一起构成了具有稳定性的中国优化共生框架。

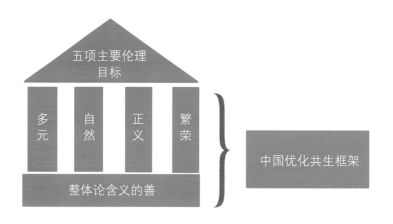

中国优化共生设计方案（COSDP）图示

1. 中国优化共生框架的善

在现代科技世界，人们越来越多地遭遇多元古代传统和现代人类价值的冲突。中国优化共生设计方案力图以世界良性秩序的善为

all things within their environing contexts, a shared diversity as the aspiration to maximize difference in an achieved harmony, and the always provisional, emergent nature of natural, social, and political order without any fixity or finality. The Chinese Optimizing Symbiosis Design Programme is developed as a possible reference both within China and elsewhere.

The Chinese Optimizing Symbiosis Design Programme contains three different layers from the holistic concept of the good through four meta-dimensions to five principal objectives. Both the holistic good and the meta-dimensions together constitute the fundamental and stable framework of the programme.

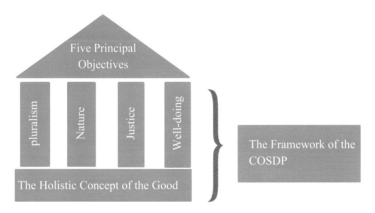

The Chinese Optimizing Symbiosis Design Programme (COSDP)

1. The Good in the Framework of the COSDP

In the modern scientific and technological world, people are experiencing an increasing degree of tension both among a diverse range of classical traditions themselves, and the relationship that these traditions have to contemporary human values. The Chinese Optimizing Symbiosis Design Programme aims to integrate these diverse traditions and contemporary values on the basis of a

基础来整合这些不同的伦理传统。[21]与此同时，机器人技术的广泛使用也在深刻影响着自然环境和自然物种多样性。整体论含义的善，需要从历史的视角来关注自然和文化的价值多元结构。机器人系统的设计和研发应该允许并保护世界多元价值的共生秩序。

在人类文明历史中，对于技术和技术品的偏见一度根深蒂固。长期以来，技术品被狭隘地界定为执行人类意图和实现人类目标的单纯工具。技术品被错误地认为只具有工具性，因而对于价值保持中立，或者只能从人类的意图和目标中获得附加价值。随着现代科技的迅猛发展，我们长期形成的对于技术品的偏见需要彻底摈除。[22]《美国机器人业路线图》（2016）指出"我们距离多种多样地拥有全面和通用自主功能的设备和手段还有10年到15年的时间"[23]。作为半自主和全自主智能系统，机器人将通过与人以及其他环境因素的协作来彻底重建人类社会生活结构以及人类与自然界的共存关系。机器人系统将与人类和其他因素一起成为建制性要素，贡献、分享和展示目标机制的核心价值或价值图景。在中国古典传统中，伦理价值并不基于个别意图、行动和结果的道德评价，而是依赖个人生命的整体统一。个人生命的道德评价要依赖她（或他）在生活共同体的良性秩序中所扮演的角色和所发挥的功能。借鉴中国传统的道德评价依据，机器人系统的设计和研发应该明确考虑该系统在建立、维护和促进目标机制核心价值或价值体系中所要扮演的角色，以及为实现该角色所必须整合的能力。[24]

holistic conception of the good for a well-ordered, integrated world.[21] On the other hand, the wide deployment of robotic technologies makes huge impacts upon biodiversity and natural environment. From the viewpoint of the holistic good, we are required to do justice to the natural and cultural plurality with a historical dimension. It is thus proposed that the design and deployment of robot systems should be aligned with a well-ordered, integrated world.

A biased view about technology and technological artefacts has prevailed in China over the millennia. Technological artefacts have too often been misunderstood as instruments used to satisfy the intentions of people and to achieve their goals. The use of technological artefacts has thus been seen to be indifferent to established values, or at best having only values relevant to our human intentions and goals. Nowadays this prejudice against technological artefacts must change because of the unprecedented progress being achieved in modern science and technology.[22] It is acknowledged in the *Roadmap for US Robotics* (2016) that "we are still 10 to 15 years away from a wide variety of applications and solutions incorporating full-scale, general autonomous functionality."[23] The robot as a sort of semi-autonomous or autonomous system may collaborate with humans and other relevant components to structure our social lives and to coexist with non-human entities. Together with humans and other elements, robot systems as a constitutive component may function to contribute to, share in, and manifest the core value or values of relevant institutions. In the Chinese classical traditions, moral value is not based on the evaluation of particular intentions, actions, or consequences, but rather on the unity and quality of the whole life narrative. The moral value of individual persons must be measured against the role that he or she should play for the sustained harmony of the whole community. The design and development of robot systems should explicitly take into account the capabilities that the role of robots might provide in articulating the core values as they exist within relevant institutions.[24]

个体自主、权利平等以及参与立法，这些现代价值肇始于近代欧洲并获得今天世界各国的普遍认可。经过一个多世纪的发展，中国今天早已成为现代化的国家。尽管中国的机器人伦理标准化方案深刻借鉴了中国古典传统，中国优化共生框架同样致力于在机器人系统中彰显全球普遍认可的现代人类价值。这些宝贵的现代人类价值不仅不该被拒斥或边缘化，而且应该成为我们今天世界良性秩序不可或缺的一部分。

过去几年，机器人技术在生产制造、陪护应用、无人驾驶汽车以及飞行器领域获得了长足进展。拍照系统、成像、通信系统以及基础运算这些核心领域也日新月异。[25] 装配自主系统的机器人可以被广泛用来探索太空以及我们星球上迄今为止人类尚未涉足的空间领域。历史经验表明，对于未知地域的探测开发必然会创造经济增长，增加一国的资源、知识和力量。然而，户外机器人的无节制使用不可避免地会引发环境风险和伦理忧虑。人类利用自然资源能力的增强或许会导致整个星球的过度人化和过度开发。为此，人类或将支付高昂的代价，例如物种多样性的丧失以及其他形态自然物的消失。另外，生物机电一体化以及移植手术领域的技术进步带来的新的生命医学伦理问题，导致我们对于自己身体完整性和本属价值的伦理忧虑。绝大多数中国传统具有对生态的深厚情感并且明确尊重自然界中超出人类控制或利用的内在价值。为此，中国机器人的设计和研发要保护自然的多样性，不仅出于人类自身可持续发展的目的，而且为了自然本身的内在价值。今天全世界包括中国在内多元文化

Individual autonomy, a regimen of shared human rights, and full participation in governance are modern values that have their origin in the cultures of modern Europe. Despite its explicit reference to Chinese classical traditions, the Chinese Optimizing Symbiosis Design Programme is likewise committed to these modern values. Modern values should not be marginalized or rejected, but on the contrary, should be fully integrated as core components contributing to a well-ordered world.

Over the last few years, we have seen tremendous progress in robot technology, in manufacturing, healthcare applications, autonomous cars and drones, as well as in core technologies such as camera systems, communication and display system, and in basic computing.[25] Robots incorporating autonomous systems can be mobilized to explore those territories on the planet that so far have been inaccessible to us. The exploration of these unexplored territories will certainly generate economic growth and further national resources, knowledge and power. But at the same time, environmental concerns inevitably arise with the unrestrained use of outdoor robots. Increased efficiency in exploiting natural resources could result in an excessive anthropization of the planet at the expense of losses within nature's biodiversity and other forms of natural life. Advances in bio-mechatronics and in the field of implantations raise important bioethical concerns with respect to the integrity and the proper value of the human body. Most Chinese classical traditions are resolutely ecological in their sensibilities, and evidence a deep respect for the intrinsic value of nature beyond any human control or exploitation. Robotic design and deployment must be committed to natural diversity not only for the sustainable development of the human world, but also for the sake of nature itself. And because different cultures flourish both within Chinese society and across the globe,

普遍共存，这已经是不争的事实。中国机器人业只有充分尊重文化多样性才能获得真正的成功。

2. 中国优化共生框架的元向度

中国优化共生框架基于良性世界秩序的整体论框架。然而，良性世界秩序并不因此成为空洞的语词。作为具有整合力的框架，良性世界秩序所要求的整体论的善必须被具体化为四重元向度：**多元、自然、正义和繁荣**。[26]

1）多元不仅是具有整体含义的善的逻辑要求，也是我们人类生活中的普遍基础经验。多元不仅涉及文化多元而且涉及自然存在物的多元。

2）在中国文化传统中，自然一方面指自然存在和自然形态，另一方面也指超出人类意图和控制的自发性、偶然性以及有限的可控性。另外，自然概念也往往蕴涵着对于整体性秩序的要求。

3）正义主要涉及的是分配正义及其实现。共生秩序要求根据每一个部分在整体中的适配角色来分配其在自然和文化世界中所需要的条件和资源。

4）世界秩序良性与否依赖每一个有机构成部分自身的发展状态。因此，每一个构成部分的生长和繁荣是良性世界秩序的必然要求。

基于整体的逻辑要求和人类基础经验事实，上述四重元向度和整体论的善一起构成了共生性整体秩序中保持稳定的基础性框架。中国优化共生框架并非具体推导机器人伦理细则的初始原理，而是

robotic design can never expect any success without respect for cultural diversity as well.

2. The Meta-Dimensions in the Framework of the COSDP

The Chinese Optimizing Symbiosis Design Programme provides a holistic framework for the well-ordered and integrated world. Such holism does not render the notion of the good for a well-ordered and integrated world an empty term. Within an integrative framework, the holistic concept of the good must be concretized in the following four principal dimensions: **pluralism, nature, justice and well-doing**.[26]

1) The dimension of pluralism not only logically derives from the holistic concept of the good but also reflects our fundamental experience of life. Pluralism is meant to incorporate both cultural and natural plurality.

2) In the Chinese traditions, the concept of nature on the one hand can mean natural existence, and on the other spontaneity, contingency and beyond human intentions and control. In addition, the Chinese concept of nature may implicate a required order of symbiotic unity.

3) The dimension of justice deals with the problem of distributive justice and its realization. A symbiotic order necessitates that each component of the whole should be attributed its conditions and sources according to its role assumed in the natural and cultural world.

4) The well-doing of a symbiotic structure is thus dependent on flourishing of each component within.

Acknowledging the logical implication of ethical holism and our respect for biodiversity, these four meta-dimensions constitute the stable and persistent fundamental framework. Such ethical framework does not consist in deductive principles for derivation of particular regulations. The four meta-dimensions are rather to be understood as a compass to orient us in particular contexts where

像指南针那样引导我们在涉及机器人的目标语境中确定价值秩序。

我们相信，只有当机器人业的技术和创新帮助实现良性世界秩序的共生和谐，中国的机器人业才能获得最大的成功。在机器人革命的时代，中国机器人业的伦理监管将支撑我们蓝色星球上最美丽的生活。

3.2　中国优化共生设计方案的主要伦理目标

在现代科技文明的世界中，四重元向度所要确立的具体伦理目标一方面要求获得现代科学发展的理论支持，另一方面要求获得尽可能广泛的跨文化共识。同时，我们不能因此把中国优化共生设计方案扭曲为一套固定不变的僵死规定，而是始终保持对相关语境的特殊性和伦理规范普遍性的敏感。

今天，各国的机器人设计和研发的参与者逐渐形成关于机器人伦理问题的一些国际性共识。根据中国优化共生框架，这些具有共识性的问题可以被分别归纳在五个具体的伦理目标中：**人权、责任、透明度、防范滥用、共生繁荣**。[27] 这部《前瞻》无法预测和穷尽所有未来可能出现的机器人伦理问题。根据既有的国际人权公约以及全球机器人业参与者的既有共识，我们通过这五个机器人伦理目标来在当代语境下具体化多元、自然、正义和繁荣这四重元向度。这五个具体伦理目标中的每一个目标都不仅体现着共生秩序元向度的要求，而且通过元向度获得必要的理论支持。另外，这些具体伦理目标并非简单地彼此平列，而是必须在机器人所涉及的特定目标环境中获得明确的价值序列并形成完整的价值图景。

robots may be deployed.

We believe that robotics can only achieve optimal success when its technology and innovation helps to realize symbiotic harmony in a well-ordered cosmos. The ethical overseeing of Chinese robotics will conduce to a flourishing world in our ear of the robot revolution.

3.2 The Principal Objectives

In the modern scientific and technological civilization, the well-ordered world must be explicated on the one hand with the support of modern science. On the other hand, it must be based on as much inter-cultural consensus as it is possible to achieve. The Chinese Optimizing Symbiosis Design Programme should never be reduced to a single set of fixed stipulations, but must always be applied with sensitivity to both contextual particularity and normative universality.

Today there is a growing international consensus among stakeholders involved in the design and deployment of robots with respect to the following five general principles: **human rights, accountability, transparency, awareness of misuse, and shared flourishing.**[27] Here this document does not pretend to be exhaustive regarding all of the possible ethical principles of robotic design and development. The following principles are proposed as a possible concretization of the Chinese Optimizing Symbiosis Design Programme within the present world on the basis of international conventions on rights and other ethical concerns common to the international stakeholders involved in robotics. Each of the five principal objectives not only is reflecting but also justified in the four meta-dimensions above. The principles are meant to stand parallel with each other, but must be integrated into a particular moral pattern in individual contexts.

伦理目标 1：人的尊严与人权[28]

1）尊重人的尊严和人权始终是中国优化共生设计方案的最重要目标之一。

2）个人、公司、专业机构、研究所、政府以及其他中国机器人业参与者必须根据我国政府承认的国际人权公约和国际相关专业机构文件来评估机器人在人权方面的影响。[29]

3）任何时候，机器人绝不允许设计成为单一或主要以杀害和伤害人类为目的。

4）在规定运行时间内，机器人系统的安全对于人权保护具有首要意义。

5）机器人系统必须保障个人定义进入个人数据信息的权利并保障对个人数据信息的知情同意权。

6）作为工业制品，机器人不应该被设计为具有可能会导致伦理伤害的操控、诱导和欺骗功能。

7）机器人的设计必须保障潜在用户不因使用机器人而遭受歧视，或者被迫获取和使用机器人。

8）在可预见的未来，机器人只允许被设计和用作为目标机制中影响道德行动和结果的道德参与要素。[30]不允许机器人被赋予与人类责任和权利同等的责任、权利以及优先权。

伦理目标 2：追责[31]

1）基于文化环境、应用和机器人用途，个人和机构需要不同程

Principle 1: Human Dignity and Human Rights[28]

1) Respect for human rights and human dignity is a core goal of the Chinese Optimizing Symbiosis Design Programme.

2) Individuals, companies, professional bodies, research institutions, governmental agencies, and all relevant others are required to assess the human rights impacts of robotics in accordance with international human rights instruments of UN and other relevant documents of international professional institutions which Chinese government admits.[29]

3) Robots should never be designed solely or primarily to kill or harm humans.

4) The safety and security of robots throughout their operational lifetime is of the foremost importance in protecting human rights.

5) Mechanisms are required to protect the right that people have to define access and to provide informed consent with respect to their personal data.

6) Robots as manufactured artefacts should not be designed to cause likely ethical harm through manipulation, coercion, or deception.

7) Potential users should not be discriminated against, or forced to acquire and use a robot.

8) For the foreseeable future, robots should only be designed and used as moral factors to influence moral actions and outcomes within the network of the target institution.[30] Robots should not be granted responsibilities, rights and privileges equal to human responsibilities and rights.

Principle 2: Accountability[31]

1) Given the cultural context and the application and use of robots, people and institution need various levels of clarity with respect to the manufacture and deployment of robot systems to establish accountability and to avoid

度地清楚了解机器人系统的制造和运行，以便确立责任归属并规避潜在伤害。

2）在任何情况下，设计者和制造者都始终对他们的系统造成的风险或外部影响承担责任。机器人设计者和制造者必须能够在程序层面说明为什么系统以某种方式运行，以便应对和处理责任归属问题，或者在多重设计者、制造者、拥有者以及操作者之间进行责任分配，避免在公众间引发混淆和恐惧。

3）及时建立由不同的相关参与者共同构成的机器人业态系统，帮助制定相应规范。

4）必须建立注册和档案系统，帮助我们在任何时候都可以查询到特定机器人的法律责任承担主体。机器人生产者、操作者和拥有者必须注册密码和高等级参数（包括目标用途、训练数据和环境、传感器和现实世界数据源、算法、图像处理、型号特征、用户界面、执行器和输出，以及优化目标、功能丧失和奖励功能等）。

5）今天，很多与责任归属和检验相关的具体设计指南还无法在技术上实现。我们建议这类责任相关的伦理问题通过采取明确告知的方式来解决。

伦理目标3：透明度[32]

1）机器人系统的透明度意味着它的可回溯性、说明性和解释性。机器人系统要确保人们总能够发现系统如何以及为什么以特定方式做出特定决策和进行特定行动。

potential harm.

2) Designers and manufacturers must remain accountable for the risks and collateral damage their system cause. Designers and manufacturers of robot systems, to avoid confusion or fear within the general public, must be able to provide the programmatic-level accountability proving why a system operates in certain ways that can be appealed to in addressing issues of culpability necessary to properly apportion liability among several responsible designers, manufacturers, owners, and/or operators.

3) Multi-stakeholder ecosystems should be developed to help create norms where they do not exit.

4) Systems for registration and record-keeping should be created so that it is always possible to find out who is legally responsible for a particular robot. Manufacturers/operators/owners of robots should register key, high-level parameters incorporating intended use, training data/training environment, sensors/real world data sources, algorithms, process graphs, model features, user interfaces, actuators/outputs and optimization goal/loss function/reward function.

5) Many of the desired design specifications regarding accountability and verifiability are not yet technologically possible. The ethical issue of accountability is best addressed through full disclosure.

Principle 3: Transparency[32]

1) The transparency of a robot system consists in its traceability, explicability and interpretability. Such information is necessary to discover how and why a robot system makes a particular decision and acts the way it does.

2）在性能鉴定中，机器人系统能够开放和展示系统进程和输入数据。

3）在用户使用中，机器人系统能够以简洁的方式帮助用户理解系统正在做什么以及为什么要做。

4）当事故发生时，进行案件调查的司法人员、律师和专家取证人员能够获得关于机器人系统必需的证据和决策。

5）具有可能扰乱社会和严重安全风险的机器人技术（例如无人驾驶汽车和手术机器人），必须对较大范围的公众保持一定程度的透明度，从而建立公众对于技术的信心，促进安全操作并帮助赢得更广泛的社会认可。

6）机器人业参与者必须共同制定机器人系统透明性测量和检验等级的规范，确保机器人系统能够被客观评估以及按照等级被客观界定。

伦理目标4：滥用及防范意识[33]

1）机器人系统必须按照既定的设计方式、设计目标和设计领域使用。机器人系统必须能够避免或者发出信号提示任何可能形式的滥用。

2）相比于传统科技产品，机器人技术会导致更大的滥用风险，例如黑客、误用私人数据、赌博或者利用。机器人业的参与者必须广泛和有效地提供伦理教育并帮助公众建立自我保护意识，让全社会警惕机器人滥用的潜在风险。

2) For validation, transparency makes the system's processes and input data available for scrutiny.

3) For users, transparency provides a simple way for them to understand what the system is doing and why.

4) Following an accident, judges, lawyers, and expert witnesses involved in the trial process require transparency that informs both the evidence and the decision-making.

5) For disruptive and safety-critical technologies (such as driverless cars and surgery robots), a certain level of transparency for society as a whole is required to build public confidence in the technology, to promote safer practices, and to facilitate wider societal adoption.

6) It is necessary to develop new standards that can describe measurable and testable levels of transparency, so that systems can be objectively assessed and levels of compliance determined.

Principle 4: Misuse and Awareness of Misuse[33]

1) A robot system should always be used in its designed ways, for its designed purposes, and in its designed fields of operation. The system must be designed to avoid or signal any forms of potential misuse.

2) Robotic technology gives rise to greater risk of misuse through activities such as hacking, the misuse of personal data, gaming, and exploitation. It is necessary in scalable and effective ways to provide education on ethics and security awareness that make society sensitive to the potential risks of robotic misuse.

3）尤其要对政府、立法机构和执法机构就机器人滥用的危险、伤害和风险进行必需的相关教育，特别要防范执法人员在执法过程中对公众造成的恐慌。

伦理目标 5：共生繁荣优先 [34]

1）在机器人设计、研发和使用中，仅仅规避非意图负面后果和增加效能、产能和利润是根本不够的。离开对于共生秩序的考虑，机器人仍然会对个人心理健康、自尊、情感、获取目标能力、环境等造成严重负面后果。[35]

2）机器人系统必须能够帮助展现目标机制的共生整体价值。除了经济繁荣以外，良性秩序还要涉及人类安康和自然（人类身体和非人类自然物）的内在独立价值。在人的尊严和人权以外，良性秩序应该涉及个人自主、文化多元（个体、社群和政府治理的条件差异）、人类能力、公平劳动以及自然多样性。

3）机器人设计和研发应该完整测量人类生活质量，建议综合考虑主观性、客观性、复合型指数以及社交媒体数据。人类安康的综合指标体系确立应该基于广泛讨论。关于人类安康的综合指标体系不应该被规定为一套僵化不变的指数，而是能够根据文化环境差异进行相应调整。

4）一旦人类安康的综合指标体系被社会接受和承认，那么所有相关指标都应该能够翻译并整合为机器人系统的次级功能目标。

3) It is necessary to educate government, lawmakers and enforcement agencies about the potential hazards, harm and risks of robotic misuse.

Principle 5: Prioritizing Shared Flourishing[34]

1) It is not sufficient in increasing efficiency, productivity and profit in the robotic design and use to merely avoid negative, unintended consequences. Without considering the well-ordered environment, the socially responsible design of robots can still produce dramatic negative consequences for people's mental health, their self-esteem, their emotional stability, their ability to achieve their goals, the quality of their environment and so on. [35]

2) Robot systems must be able to contribute to articulating the core values of the well-ordered environment in a target institution. The well-ordered world should not only be limited to the well-doing of human beings (apart from economic and financial prosperity) but also the intrinsic value of nature (both the nature of the human body and non-human nature). Apart from human rights and human dignity, the good of the well-ordered world should incorporate individual autonomy, cultural diversity (including individual, social and governmental circumstances), capabilities, fair labour as well as natural diversity.

3) For a complete measurement of the quality of life, it is recommended to take into account subjective, objective, and composite indicators, as well as social media sourced data. A comprehensive metrics of people's well-doing should be developed on the basis of the widest possible consideration. The metrics of well-doing should not be stipulated as a set of rigid indicators, but should always entail flexibility and a sensitivity to cultural diversity.

4) Once such metrics are acknowledged as a directional driver for society, the existing measures should be translatable into sub-objectives to be incorporated into the functions of robot systems.

5）人类安康的综合指标体系和激励也应该考虑并结合就促进、保护和实现人权的国际义务方面所做的第三方评估。

6）机器人系统的设计应该避免反人性化和不恰当操控人的选择。

7）人类身体是人格的构成条件、价值的承载者和社会规范的基础。机器人系统的设计和研发不允许损害个人身体、他人、人类价值以及公共利益。

8）基因技术和机器人技术结合对于人类安康造成的潜在影响迄今尚不明确。我们建议尽快成立由基因技术、机器人学、心理学、伦理学和政府人员共同构成的委员会，开始了解和研究运用机器人技术解读基因数据会对人类安康造成的影响。[36]

9）机器人业标准必须包含维护环境可持续性的相关措施，彰显对生态环境的责任并确保环境正义。

10）在机器人系统撤出使用之前，需要根据中国政府承认的国际人权保障公约或专业文件和人类安康的综合指标体系，对于机器人系统撤出使用造成的影响进行第三方评估。

3.3 中国优化共生设计方案的实施方法

中国优化共生设计方案不应被等同为一套初始原理，用来推导机器人设计和研发所要实现的具体伦理规范。机器人功能所要展现的价值图景必须根据它应用的具体语境来建立。

中国优化共生设计方案提出的是一个**前瞻性**伦理方案。这个伦

5) Well-doing metrics and mechanism should also take into consideration and happen in conjunction with a third-party assessments on respect for, and international obligations to promote, protect and fulfil the full spectrum of human rights.

6) Robot systems should be designed to avoid dehumanization and inappropriate control over human choice.

7) The human body is both the ground of social rules and a carrier of values. Robot systems should not be designed and developed to harm people, or to compromise human values and common goods.

8) It is not yet clear how much impact the convergence of genomics technology and robotics will have on human well-doing. It is recommended that a working committee comprised of those at the cutting-edge of genomics, robotics, psychology, ethics and governance be convened to initiate a conversation among the different communities to better understand the impact that use of robotics to interpret genomics data will have on human well-doing.[36]

9) Guideline for robotics should include measures for ecological and environmental sustainability that mandate responsibility for the biosphere and ensure equitable environmental justice.

10) Before a withdrawal of the use of robots, it is necessary to assess the potential impacts of such a withdrawal on individual, society and natural environment by the third party in accordance with international human rights instruments of UN and other relevant documents of international professional institutions which Chinese government admits, as well as the acknowledged well-doing metrics.

3.3 A Methodology for implementing the Chinese Optimizing Symbiosis Design Programme (COSDP)

Neither the framework of the well-ordered world nor its current concretization should be considered as providing the deductive principles for the design and development of robots. The structuring of guidelines or codes for technological

理方案鼓励和要求在机器人的功能设计中展示、维持和促进目标使用机制的核心价值系统。尽管中国优化共生设计方案要推动中国机器人设计和研发与伦理相一致，但是这并不要求机器人学家自己成为伦理领域的专家，而是倡导机器人学家和伦理学家之间的密切合作。我们提议，在机器人系统的设计和研发全部进程中配置和整合伦理设计组，以此确保生产出来的机器人能够规避所谓的"第一代问题"。

在机器人设计和研发的全部进程中，机器人功能所展示的价值图景必须依赖**跨学科研究、开放性公共讨论**以及**反思平衡式的动态伦理慎思**。为了更加有效地推动中国优化共生设计方案的实施，我们设计了一个包含五步骤的机器人价值基设计程序。中国优化共生设计方案的五步骤程序既可以用来评估和改进既有的机器人系统，也可以用来引导和推动与良性世界秩序一致的机器人技术创新。

中国优化共生设计方案的五步骤程序既需要伦理学家来引导和组织，同时也需要与所有机器人设计和研发的参与者共同协作。其中，关键要素是鉴定机器人使用目标机制的价值图景及机器人所承担的功能角色，并把机器人所需要维系和促进的价值要素转译为机器人学家可以理解的功能性指南或代码。我们的五步骤进程简单描述如下：[37]

1）基于中国优化共生框架和伦理目标，鉴定机器人应用的目标机制的伦理评价空间和价值图景。

2）在引入机器人之前，对于目标机构的核心价值以及价值优先

incorporation into the functions of robots must always be made in the particular context for which robots are designed.

The Chinese Optimizing Symbiosis Design Programme proposes a *prospective* and emergent ethical programme. It encourages and demands technological articulation of robotic capabilities consistent with the core values of the relevant institutions. Although the Chinese Optimizing Symbiosis Design Programme promotes ethically aligned robotic design, it does not require that roboticists become specialists in ethics, but instead recommends close collaboration of roboticists and philosophers of ethics. This ethical sector must be integrated as an indispensable component throughout the technological processes of robotic design and development to prevent "the first generation problem" wherein failures are only corrected after they occur.

For such structuring, *inter-disciplinary research*, *open public discussion* and *reflection-balanced deliberation* are all required throughout the entire process of robotic design and development. To promote an efficient use of the Chinese Optimizing Symbiosis Design Programme, we have developed and recommend a five-step process of a value-based design methodology. The methodology can not only be applied to evaluate and improve the existing robotic design but can also guide and promote prospective innovations that will benefit the well-ordered world.

The five-step process must be initiated and organized by philosophical ethicists in collaboration with all of the other stakeholders involved in the design of robots. It is of crucial importance on the one hand to identify the moral landscape and the role which robots are designed to assume, and on the other to translate the values that the deployment of robots is meant to promote into guidelines or codes accessible to roboticists. The general description of the five-step process is as follows.[37]

1) An interdisciplinary identification of the evaluative space and the moral landscape for the target institution on the basis of the above framework and

性进行阐释，并和所有目标机制参与者进行讨论。根据讨论结果，对于目标机制价值图景进行调整和补充。

3）伦理学家与机器人设计人员讨论后者所设想的机器人角色、机器人功能范围、机器人职责以及由此带来的价值图景的可能偏移。伦理学家把机器人所要辅助展现的价值转译为机器人系统所必须具备的功能指南或代码。

4）在技术力量的帮助下，把机器人投入到目标机制中，并向用户说明如何使用机器人、机器人的能力以及局限。

5）在机器人运行一段时间后，伦理学家需要再次与用户进行讨论，了解目标机制工作流程中的变化。在必要情况下，伦理学家需要和用户一起分析人机协作中带来的一些新想法或者冲突。

作为价值的参与因素，机器人的装配要协助和推动目标机制（例如健康护理）的核心价值图景。通过上述五步骤程序，被鉴定出的价值图景应该能够展现在机器人的不同功能的组合整体中。机器人要通过所有功能的整理来实现自己在目标机制中被交予的角色和职责。中国优化共生设计方案的核心使命就是要推动伦理价值在机器人系统功能中的体现，并促进中国机器人设计和研发的伦理关怀和社会责任担当。

principles is required.

2) Interpretations of the core values and their prioritization prior to the introduction of robot design are to be outlined for discussion with all the stakeholders who are directly or indirectly involved in the institution.

3) The philosophical ethicist should participate in a discussion with designers on what role the robot will assume, and according to the vision of the design team, what the robot will be capable of doing, what responsibility the robot will be delegated, and how the core values might be affected. The ethicist should then be able to translate the identified moral values that the robot is designed to assist with and manifest, into functional guidelines incorporated into the system.

4) The robot can then with the assistance of technical support be placed into the target field, indicating how to use the robot, what it is capable of and what its limitations are.

5) After the deployment of the robot for a short period of time, the philosophical ethicist meets once again with the staff to discuss the new pattern of activities and to work out any unresolved concerns or conflicts in the use of the robot.

As a constitutive moral factor, the robot is equipped to assist and to promote the core values of an intended institution (for example, healthcare). Through the five-step methodology, the moral values should be identified and articulated in diverse functions that together constitute the robotic capabilities for its intended role in the target institution. The technological articulation of moral values through the functions of robot systems is the major goal of the Chinese Optimizing Symbiosis Design Programme to encourage and promote the socially responsible and ethically driven design and development of robots.

第四章 机器人伦理风险评估与应对

Chapter 4　Ethical Risk Assessment and Mitigation

机器人伦理风险的对象必须是具备物理形体且具有运动能力的半自主或自主智能设备。机器人伦理风险不仅包含机器人已经造成的不良伦理后果，也包含可预测的机器人技术发展将会带来的前瞻性伦理影响。基于"中国共生优化设计方案"并结合现有机器人发展状况以及人机交互的未来技术前景，我们以理性前瞻、动态调整和多元平衡的模式设计并制定机器人伦理风险框架系统。这套伦理风险框架系统要为"中国共生优化设计方案"力图达成的伦理目标提供检验标准和应用场景基础。借助跨学科研究以及国际机构对于机器人风险的既有评估，我们对机器人伦理隐患、伤害以及风险的内容条目与影响范围进行提示性分析，并给出相关风险的概览性指南。

虽然机器人的设计和研发要基于用户中心，但是中国共生优化设计方案同时凸显人权与共生繁荣的根本价值目标。责任以及透明性是确保中国机器人伦理目标实现的必要条件。避免滥用和误用则倡导国家以及全社会在政策、法规、行业规范、个人行为等各个层面形成一个有机的伦理监管体系，有效推动机器人和谐地融入自然和人类社会。

第四章　机器人伦理风险评估与应对

The object of robotic ethical risk must be a semi-autonomous or autonomous smart device with a physical body and the relevant mobility. The ethical risk of robots includes not only the adverse ethical consequences that the robots have caused, but also the predictable ethical impact that robotic technology development will bring about. Based on the "Chinese Optimizing Symbiosis Design Programme" (or COSDP) in conjunction with the development status of the existing robots and technology prospects of human-robot interaction now and in the future, we have designed and developed a system for robotic ethical risks assessment that constitutes a rationally predictable, dynamically adjustable, and dialectically balanced model. This set of ethical risk system should provide the standards for investigations and applications that confirm to the ethical goals set by COSDP. With interdisciplinary research and the existing assessments of robotic ethical risks in the relevant areas as outlined by international institutions, we provide an analysis of the content as well as the scope of the ethical hazards, harms, and risks of robots, and briefly suggest some general mitigations.

Although it sustains user-centered robot design and R&D, the COSDP underscores the fundamental ethical goals of human rights as well as symbiotic prosperity. Responsibility and informational transparency are necessary conditions to ensure the realization of the Chinese robotic ethical goals. To avoid (potential) misuse and abuse, it advocates that the government and the whole society should form an interactive supervision system that is intended to be sensitive to the relevant ethical risks at various levels. The relevant observation would, in turn, produce a series of actions, including the making of policy, the setting of regulation, and the constructing of norms, that provide us with guidelines for the relevant individual and industry-wide conduct. This would help promote the harmonious integration of robots into nature and human society in a more efficient way.

4.1 机器人伦理风险评估范围

机器人伦理风险框架的评估对象范围包括，在人机交互的可能场景中直接或者间接被影响到的人类个体、以各种互动方式组织起来的人类社群、其他生命物种以及相关联的自然环境。在机器人伦理风险的具体评估中，实际参与评估与讨论的主体不限于机器人研究和设计者、伦理学家、相关社会科学领域的专家、机器人用户及其家庭成员，还需要尽可能广泛地包含涵括具备适当信息知情条件的公众以及对于机器人伦理讨论具有明确兴趣的社会成员。在对机器人伦理风险与评估进行有效定性或定量过程中，评估人员与机构进行数据信息采集和提取工作需要严格遵循相关法律法规以及行业准则。评估人员和机构必须在尊重和确保个体隐私的前提下，保持信息数据搜集的公平与公正。

机器人伦理隐患、伤害以及风险评估需要考察的范围包括（但不限于）：

1）是否充分尊重个人知情同意权、隐私权等相关权利，是否充分尊重个人尊严，是否存在未经正当授权情况下的个人数据信息收集，是否存在对相关数据的误用。

2）是否充分考虑机器人设计、使用和人机交互场景的复杂性与多样性，是否充分考虑在上述条件下对个人、社群和社会产生的物理与心理层面的影响。机器人的成本－收益评估是否充分体现个体、社群、社会、自然环境的权益权重。

3）是否合理地鉴定出机器人设计、使用和人机交互过程中安全

4.1 The Scope of the Robotic Ethical Risk

The scope of the assessment of the robotic ethical risk covers the human individuals that are directly or indirectly affected in possible scenarios of human-robot interaction, of human communities organized in various ways, of other living species, and of associated natural environments. In the specific and concrete assessment of robotic ethical risks, the relevant participants in the discussion are not limited to robotic researchers and designers, philosophical ethicists, the relevant experts in related social sciences, robot users and their family members. The public should also be involved: the well-informed members of the community who have a clear interest in robotic ethics discussions should also have their voices heard. In the process of effectively qualifying or quantifying the ethical risks concerning robotics, the persons and institutions participating in the evaluation need to comply with relevant laws, regulations, and industry guidelines for data collection and extraction. Evaluators and organizations must maintain the fairness and justice of the collection of information data on the premise of respecting and ensuring individual privacy.

The scope of robotic ethics hazards, harms, and risk assessment that need to be examined includes (but is not limited to):

1) Whether the relevant rights such as well-informed consent and confidential privacy are fully respected, whether human dignity is fully respected, whether there is certain data collection without proper authorization, and whether there is a misuse or abuse of related data.

2) Whether the complexity and diversity of robotic design, use, and human-robot interactive scenarios are fully considered, and whether the physical and psychological aspects of individuals, communities, and society under the above conditions are fully considered. Whether the robot's cost-benefit assessment fully reflects the rights of the individual, the community, the society and the natural environment.

3) Whether the relevant obligations, liabilities, and obligation or responsibility attribution in the design, use, and human-robot interaction are reasonably identified and properly estimated.

性、可靠性等相关义务以及责任承担方与归属方。

4）对于特种用途机器人，在其相应的设计、使用和人机交互关系中，是否针对其特性进行特定的伦理危害与风险的鉴别和评估，是否充分进行了相应伦理价值辩护。

5）对于不同种类或者类型的机器人，在其相应的设计、使用和人机交互关系中，是否充分考虑到相关功能差异所带来的机器人阶层划分及其对于人类个体、社群与社会的派生性影响。

6）在相关伦理准则不一致甚至冲突的情况下，对于选定某一或者某些准则作为核心或者赋予高权重（或优先性）操作，是否经过审慎的伦理辩护与合理说明。[38]

7）在没有（明显）伦理准则冲突的情况下，是否充分考虑机器人应用在不同社群文化以及不同历史－自然环境中所（可能）引发的不同后果，对相关结果是否进行充分的伦理风险评估。[39]

为了清晰简洁地呈现与机器人应用相关的伦理隐患、伤害以及风险评估，也为了系统性地展示相关伦理领域的概貌，我们采用图表的形式来表述机器人伦理风险评估框架。在此之前，我们有必要对表格中纵横轴的设计初衷和范畴排序予以说明。

4.2 机器人伦理风险评估表结构说明

表格横轴依次排列考察领域或范围、伦理隐患、伦理风险与应对四项内容范畴。对于伦理隐患、伦理风险和应对这三个范畴的选择与安排，反映出中国共生优化设计方案与国际通行方案之间的接

4) Whether or not there are some specific identification and assessment of ethical hazards and risk for a robot with a special design purpose or special uses (especially its characteristics in its corresponding design, use, and human-computer interaction relationship), and whether adequate justification of the relevant ethical value is fully implemented.

5) For different kinds or types of robots, in their corresponding design, use, and human-machine interaction relationship, whether the hierarchical classification of the robot and its derivative implications for human individuals, communities and society are taken fully into account.

6) In the case of inconsistent or even conflicting ethical standards, whether the preferred priority of the criteria used in the resolution of the relevant problems is prudently justified or reasonably explained in an ethically acceptable or appraisable way.[38]

7) In the absence of (obvious) conflict of ethical standards, whether to fully consider the different consequences (possibly) caused by the application of robots in different cultural communities and in different historical-natural environments, and whether to perform the adequate ethical risk assessment on the relevant results.[39]

To clearly and concisely present the ethical hazards, injuries, and risk assessments related to robot applications, as well as to systematically show the general ethical landscape, we use a table to present the framework of ethical risk assessment for robots. Before introducing more details, we shall first explain the categories selected and located in the rows and columns.

4.2 Explication of the Table of the Robotic Ethical Risk

The rows of the table are arranged in four categories: the field or scope, ethical hazards, ethical risks, and mitigation. The selection and arrangement of the latter three categories reflect the continuity and compatibility between the COSDP and the other international projects on robotic ethics that in turn indicate acknowledged consensus in the field of robotic ethics at the international level.[40] Ethical hazards include those ethical hidden or potential

续性与相容性，同时也反映出机器人相关者在伦理评估层面的国际共识。[40] 伦理隐患主要考察机器人在目标领域可能产生的负面影响；伦理风险侧重机器人负面影响可能产生的环境和条件；应对方案仅仅表达规避和减少机器人伦理负面影响的基础原则。

表格纵轴依赖中国共生优化设计方案区分出"个人与个体""社群与社会""经济与商业"以及"环境与共生系统"四项考察对象（领域）。这四组对象领域的划分突破了人类中心主义的伦理倾向，呈现出逐渐增强的整体论特征。第一，基于中国优化共生框架，这四组对象领域的划分从机器人产品的终端使用者个人出发，认可个人身体的自然性内在价值；在尊重多元、倡导正义的主张下，凸显人类以及机器人都是自然共生系统中的成员，公平公正地认可人类使用者、机器人个体以及相关自然物的权益与价值。中国共生优化设计方案在整合现代个体价值的同时，强调个人在社群中的人际角色和教化品格的伦理特质，在社会层面进一步凸显身体的重要伦理载体意义。第二，就目前既有机器人科技而言，机器人尚不能被视为完全自主的道德行动承担者。然而，机器人所具备的道德要素特征无可否认。机器人伦理风险评估既需要在个体层面也需要在社群生活整体中来考察机器人在"目标机构"中的整合性。第三，在全球化的背景下，商业与经济既是各国机器人业发展的核心驱动力，也是社群以及国家之间互动沟通的重要组织形式。机器人在工业生产和商业系统中的整合性需要获得考察和评估。最后，环境与自然所构成的整体性共生系统是上述所有对象（或领域）及其相关活动的基

dangers or harms or some possible adverse effects that may be brought about by robot application in the targeted field; ethical risks focus on the possible contexts and conditions within which the relevant hazards occur; and mitigations generally suggest the basic principles of reducing and avoiding the negative effects of those hazards and risks.

The categories in the columns of the table, based upon COSDP, includes four subject matters or fields: "persons and individuals", "community and society", "economics and business", and "environment and symbiosis system". The division of these four fields reflects the explicit abandonment of anthropocentrism and the adoption of holism in character. First, in the framework of the symbiosis of optimizing symbiosis, the division of these four groups of objects starts from the end-users of robot products and recognizes the intrinsic value of the individual's body; under the proposition of respecting pluralism and advocating justice, it highlights humanity. And robots are members of the natural symbiosis system and reasonably recognize the rights and interests of human users, robots, and related natural objects. The Chinese symbiosis optimization design scheme emphasizes the ethical characteristics of the individual's interpersonal roles and educative character in the community while integrating the individual values of modern society, and further highlights the essential ethical meaning of the body at the social level. Second, as far as robotics technology is concerned, robots cannot yet be regarded as fully autonomous moral actors. However, the characteristics of the moral elements possessed by robots are undeniable. The ethical risk assessment of robots needs to examine the integration of robots in target institutions both at the individual level and in the overall community life. Third, in the context of globalization, business and economy are both the core drivers for the development of the robotics industry in various countries as well as significant forms of interaction between communities and nations. The integration of robots in industrial production and commercial systems needs to be reviewed and evaluated. Finally, the symbiosis system of environment and nature is the basis and platform for all subject matters (or fields) mentioned above for their related

础与平台。机器人在环境与自然中的整合性会在伦理价值风险评估当中扮演非常重要的角色。纵轴所罗列的相关对象（或领域）均具备各自的内在价值，应当享有各自相应的权利并获得充分的尊重。中国优化共生设计方案所倡导的机器人伦理体系是真正实质的多元主义方案。

机器人伦理风险评估框架能够展示中国优化共生设计方案所鉴定出的机器人伦理隐患和伦理风险地形图。在个人与个体层面，我们考察个人权利和机器人权益是否得到合理尊重以及机器人仿真和智能特征可能造成的伦理风险。在社群与社会层面，我们考察机器人对于家庭、社群、文化族群多元性可能带来的伦理风险。在经济与商业层面，我们考察机器人在工业生产和商业活动中带来的安全、就业、责任归属和知情同意等方面的伦理风险。在环境与共生系统层面，我们考察机器人运行带来的环境压力和资源浪费等伦理风险。

下面表格中所涉及和罗列的机器人伦理隐患、伤害以及风险并非完备。在机器人运行的所有领域中，机器人对个人、社会和环境造成的安全问题考虑必须贯穿始终。另外，由于机器人技术的快速发展，我们可以合理地预期，未来还会有其他新的内容被纳入伦理伤害与风险评估的考察范围之中。中国机器人伦理标准化体系由于采取世界共生秩序的整体善作为基础，能够在保持框架稳定的前提下拥有较强的可塑性和容纳度。中国优化共生设计方案是一套前瞻性的伦理体系，能够以积极的态度面向（乃至欢迎）未来可能增加的伦理风险评估内容。

activities. The integration of robots into the environment and nature shall play an essential role in the ethical value risk assessment. The relevant objects (or fields) listed on the vertical axis all have their intrinsic value and should enjoy their respective rights and be thoroughly respected. The robotic ethical system advocated by COSDP is a truly substantial pluralism program.

The system of robotic ethical risk assessment can demonstrate the topographic maps of the ethical hazards and risks of robots identified by COSDP. At the person and individual level, we examine whether human rights and robot rights are reasonably respected and the ethical risks that may arise from robot simulation and intelligence features. At the community and society level, we examine the ethical risks that robots may bring to the diversity of families, communities, and cultural communities. At the economic and commercial level, we consider the ethical risks of robotics regarding safety, employment, responsibility, and informed consent in industrial production and business activities. At the level of environmental and symbiotic systems, we examine the ethical risks of environmental pressure and the waste of resources brought about by the operation of robots.

The ethical hazards, harms, and risks of the robots involved and listed in the table below are not intended to be complete. In all areas where robots operate, robots must be consistent with personal, social, and environmental safety issues. Also, due to the rapid development of robotics technology, we can reasonably expect that there will be other new content that will be included in the scope of ethical injury and risk assessment. Because the Chinese system for the standardization of robot ethics adopts the overall good as the basis of the world symbiosis order, its theoretical system can have substantial plasticity and accommodation degree under the premise of maintaining the framework stability. The Chines Optimizing Symbiosis Design Program is a set of forward-looking ethics systems that can positively make (even welcome) ethical risk assessments that may increase in the future.

4.3 机器人伦理风险评估表

	领域或范围	伦理隐患	伦理风险	应对
个人与个体 / 个人	个人隐私和权益保护	侵害个人隐私；侵害知情权	在未经授权的情况下，搜集和公布机器人使用者的个人信息可能会造成滥用相关信息与侵犯个人隐私	通过明确规定，确保当事人的知情权，有效保护数据，严格指明相关数据的使用领域
	个体信任	使用者丧失对于机器人终端信任的危机，产生关于机器人使用的个体性或者群体性偏见以及过激行为	由于信息不明或者信息误导而可能造成对于机器人的误用与滥用	确保机器人的本质不被歪曲或掩盖，明确设计意图与设计理念，避免欺骗
	心理生理依赖	使用者形成对于机器人不当的生理－心理－交际依赖或者对于使用机器人出现成瘾性行为	由于相关的不当依赖，可能会造成使用者（或群体）丧失人类个体（或群体的）自主性；破坏和剥夺使用者个体（或群体）角色	建立并提高对于相关生理－心理成瘾与依赖问题严重性的清晰认识与重视

4.3 The Table of the Robotic Ethical Risk

		the field/scope	ethical hazards	ethical risks	mitigation
persons and individuals	persons	confidential privacy and rights protection	violation of confidential privacy; violation of the right to information	unauthorized collection and publication of confidential information of robot users may result in misuse or abuse of relevant information and infringement of personal privacy	through explicit regulations, to ensure the parties' right to know, effectively protect data, and strictly specify the use of relevant data
		trust	the users lose confidence in the robot terminal and generates individual or group prejudice and excessive behavior regarding the use of the robot	misuse and abuse of robots due to unknown information or misleading information	to ensure that the nature of the robot is not distorted or obscured, clearly design intentions and design concepts to avoid cheating
		psychological and physiological dependence	the user develops an improper physiological-psychological-interactive dependence on the robot or an addictive behavior on the use of the robot	due to over-dependence, it may cause the user (or group of users) to lose the autonomy of the individual (or group); destroy and deprive the individual (or group) of the user	to establish and to improve understanding and emphasis on the severity of related physiological-psychological addiction and dependence issues

(续表)

	领域或范围	伦理隐患	伦理风险	应对
机器人	人格（与人类形体）尊严	过度的或者不当的（在人类性格或者人类形体层面的）模仿或者拟人化设计	由于过度拟人化，可能会掩盖和欺瞒机器人的机械本质，刻意隐藏机器人对于人类性格或者形体在设计层面的模拟意图或者设计性的行为或者行动表现	避免不必要的拟人化设计，对必要拟人化的设计意图进行评估和鉴定
	应用与设计意图	违反设计意图与应用范围；造成对使用者生理－心理方面的伤害	由于缺乏充分重视，可能会造成相关的误用与滥用	清楚理解并充分尊重设计意图
	机器人权益	不当使用下造成对机器人的损害；不当的回收处理	由于对机器人权益的无视，可能会造成对机器人使用过程当中的不必要的损耗，造成资源浪费；在某些情形下（例如，仿真机器人的应用），还会造成对于人性和尊严的贬低与价值丧失	去除人类中心主义的局限，尊重机器人本身的权益与价值

(continuation)

		the field/scope	ethical hazards	ethical risks	mitigation
	robot	the dignity of personality (and the form of human bodies)	excessive or improper imitation (of human characters or the form of human bodies) or overly anthropomorphic design	due to excessive anthropomorphism, it may mask and deceive the robot's mechanical essence, deliberately hide the robot's artificial intention or design behavior or action performance on the design level of human character or body	avoid unnecessary anthropomorphic design, evaluate and identify design intent that is necessary for personification
		applications and design purposes	violation of design intent and scope of application; causing physical and psychological harm to the user	lack of due attention may result in misuse and abuse	to explicitly understand and to fully respect the design purposes
		robot rights	improper use of the robot caused damage; improper recycling	due to the ignorance of the rights and interests of robots, it may cause unnecessary waste during the use of the robot and waste of resources; in some cases (for example, the application of artificial robots), it will also cause the devaluation of human nature and dignity	to reject anthropocentrism and to respect the rights and values of the robot by itself

(续表)

	领域或范围	伦理隐患	伦理风险	应对
	自主学习和能力增强	机器人在人机互动关系中异化；危害人类使用者与环境	由于缺乏对机器人自主学习范围和应用界限的实时有效监控；无法预期和掌握机器人相关能力的发展与活动；造成人类个体、社群、社会与自然资源的滥用与危害	强化人类对于机器人自主学习和能力增强过程中的监控
社群与社会	社群角色与社会关系	造成使用者个体（或群体）对其社群与社会角色的混淆与迷惑；危害人类社群与社会建构和组织	由于频繁人机互动可能会造成人机关系与社群－社会关系的混淆以至带来不当替代，无视社会阶层的矛盾	明确和突出机器人的本质，阐明机器人的设计意图，严格机器人的使用领域

(continuation)

	the field/scope	ethical hazards	ethical risks	mitigation
	self-learning and enhancement	robot alienation in human-robot interaction relationship; to cause the harms of human users and environment	due to the lack of the effective monitoring of the scope and application of the robots' self-learning activities, to fail to anticipate or control the development and activities of robot-related capabilities; and the abuse and harm of human individuals, communities, society and natural resources.	to improve the capabilities of the effective monitoring
community and society	roles in a community and social relations	to confuse the individual (or a group of) user(s) with their community and social roles; endanger the human community and social construction or organization	due to frequent human-robot interactions, it may cause confusion between human-machine relations and community-society relations, and bring about improper substitution, ignoring the conflicts among social classes	to define and highlight the nature of the robot, to clarify the design purposes for the robot, and to strictly use the robot accordingly

(续表)

	领域或范围	伦理隐患	伦理风险	应对
	社群多样性与文化多样性	破坏社群、文化与价值的多样性；加深阶层固化；强化社会偏见	缺乏对社群、文化多样性的尊重，可能会强化价值单一化取向，巩固社会既有的不当刻板观念，对少数群体的价值取向和偏好的不当边缘化与压迫	充分尊重社群与文化多样性，在设计机器人的过程中，努力保证对于多元价值的尊重与体现
	社会保障与社会阶层	由机器人的使用、消费而造成社会阶层固化，加深社会成员的阶层分裂，加重社会福利救助体系的负担	在缺乏相关社会福利保障制度的情况下，专业服务机器人的大范围使用可能会造成对于人类雇员的取代，增加失业人口比例，造成社会福利救助体系的负担，加剧社会运行与组织成本	建立合理有效的社会福利保障制度，增强社会对于再度学习与再度教育的相关资源的保障

(continuation)

	the field/scope	ethical hazards	ethical risks	mitigation
	community and cultural diversity	destroy the diversity of communities, cultures and values; deepen the stratum solidification; strengthen social prejudice	lack of respect for community and cultural diversity may strengthen the orientation of value simplification, consolidate the existing notion of improper stereotypes, and improperly marginalize and oppress the value orientation and preference of minorities	full respect for community and cultural diversity. In the process of designing robots, efforts are made to ensure the protection of diverse values.
	social security and social class	the social class has been solidified by the use and consumption of robots, deepening the social stratification and increasing the burden on the social welfare assistance system	in the absence of relevant social welfare protection systems, the extensive use of professional service robots may result in the replacement of human employees, increase the proportion of the unemployed, cause a burden on the social welfare assistance system, and increase social operation and organization costs	to establish a reasonable and effective social welfare protection system and enhance social protection for related resources for re-training and re-education

(续表)

	领域或范围	伦理隐患	伦理风险	应对
经济与商业	雇佣与协作	冲击劳动力市场；增加失业人口数量	由于机器人的不当应用与过度配置而带来对既有劳动力市场的冲击，降低被雇佣方在就业市场中的竞争能力，造成失业问题	明确机器人使用与投放领域和规模，保障劳资双方权益，提高社会保障能力
	经济与商业活动中的责任、义务、权益	无法有效归属经济与商业活动中的责任、义务、权益相关的承担方；造成不当的商业垄断；破坏相关经济与商业活动规章	由于信息和技术不对称可能会造成商业合约的不公正，以及由此引发的违反伦理道德、法律法规等方面的不当行为	保证信息透明和信息交流渠道畅通，完善相关法规

(continuation)

	the field/scope	ethical hazards	ethical risks	mitigation
	employment and collaboration	impact on the labor market; increase the number of unemployed	due to the improper application and over-allocation of robots, it brings impact on the existing labor market, reduces the competitiveness of the working class in the job market, and causes unemployment	to restrict the field and scope of use and delivery of robots, protect the rights and interests of employers and employees, and improve social security capabilities
economics and business	responsibilities, obligations, and equity in economic and commercial activities	inability to efficiently attribute the relevant responsibilities, obligations, and rights in economic and business activities; create improper commercial monopolies, and destroy regulations concerning related economic and commercial activities	due to the information and technology asymmetry may cause unfair commercial contracts, and the resulting misconduct in violation of ethics, laws and regulations	ensure information transparency and information exchange channels, improve relevant laws and regulations

(续表)

领域或范围	伦理隐患	伦理风险	应对
商业诚信与信息透明	造成商业欺诈；不当商业盈利	由于操作、应用与监控的相关信息不够清晰与透明，造成参与相关活动的人类个体与群体对于人机互动场景下相关活动的后果与风险缺乏充分的理解和把握，无法充分判断机器人相关应用与服务是否真正满足需要，无法综合计算和平衡相关收益与风险	明确相关法规或规则，清晰指明相关操作流程，严格保障相关流程的操作流程与规范
环境与共生系统	环境资源与可持续发展	资源的过度使用；由于缺乏规划造成的资源浪费；环境污染	在各种类型的机器人生产和应用中，造成对于环境的巨大压力，在缺乏有效规划和管理的情况下，造成对于资源的掠夺性使用，产生相关的工业废弃物与污染	制定有效的环境保护发展规划，合理布局发展速度与资源消耗比例

(continuation)

	the field/scope	ethical hazards	ethical risks	mitigation
	business integrity and information transparency	caused commercial fraud; improper commercial profit	due to the lack of clear and transparent information related to operations, applications, and monitoring, human individuals and groups involved in related activities do not fully understand and grasp the consequences and risks of associated activities under the human-computer interaction scenario and are unable to determine robot-related applications and services adequately. Whether to meet the needs, unable to calculate and balance relevant benefits and risks comprehensively	identify relevant regulations or rules, clearly identify relevant operational procedures, and strictly guarantee the operational procedures and specifications of related processes
environment and ecosystem	environmental resources and sustainable development	excessive use of resources; waste of resources due to lack of planning; environmental pollution	in the production and application of various types of robots, causing great pressure on the environment, in the absence of effective planning and management, resulting in predatory use of resources, resulting in related industrial waste and pollution	formulate a practical plan for environmental protection development, reasonably allocate development speed and proportion of resource consumption

(续表)

领域或范围	伦理隐患	伦理风险	应对
生态多样性	过度侵占其他物种资源；造成物种濒危或灭绝	造成环境危害与资源耗损，进而带来环境危害与生物圈环境恶化，危及地球上的植物、动物等生命物种的多样性	充分尊重地球不同生物物种的权益，在机器人设计和程序编码当中体现对于生态多样性的保护条款
局部与全局生态系统	破坏局部与全局生态系统运作机制	造成局部与全局性生态系统紊乱	准确、综合、动态地获取生态系统信息，及时进行相关评估

(continuation)

	the field/scope	ethical hazards	ethical risks	mitigation
	ecological diversity	excessively invade other species; cause species to become endangered or extinct	cause environmental damage and resource depletion, which in turn cause ecological hazards and deterioration of the biosphere environment, endangering the diversity of living species such as plants and animals on the earth	fully respect the rights and interests of different biological species on Earth, and implant code of the protection of ecological diversity into robotic design and program coding
	local and global ecosystems	to interfere operating mechanisms and to destruct of local and global ecosystem	to cause disorders of the local and global ecosystem	accurate, comprehensive and dynamic access to ecosystem information and timely assessment

第五章　护理机器人的伦理考量
Chapter 5　Ethical Reflections on Care Robots

《中国机器人伦理标准化前瞻（2019）》最后一章以护理机器人为焦点，集中分析机器人在医疗、护理、康复、助残、养老等领域所带来的伦理挑战，并尝试在设计、应用和监管层面给出应对这些挑战的初步建议。

选择护理机器人作为机器人伦理研究的范例，首先是基于现实需要的考虑。据估计，全球有超过 1/5 的人口存在运动、认知和感觉障碍，在不同程度上需要医疗和护理资源的协助。人口老龄化的加速无疑将进一步扩大需要医疗保健的人群所占人口比例。根据联合国发布的《世界人口展望：2017 修订版》统计，2017 年，全球 60 岁以上人口约 9.62 亿，占全球人口 13%，且每年以 3% 左右的速度增长。据国务院 2017 年印发的《"十三五"国家老龄事业发展和养老体系建设规划》，预计到 2020 年，全国 60 岁以上老年人口将增加到 2.55 亿左右，占总人口比重提升到 17.8% 左右。医疗卫生费用占 GDP 比重和政府支出比重也呈现出相应的增长趋势。根据世界银行的统计，2014 年全世界平均卫生费用支出占 GDP 比重为 9.9%，我国 2016 年卫生费用支出近 7000 亿美元，占 GDP 的 6.2%。我国医疗卫生市场巨大，而且仍然有很大的上升空间。然而，财政投入的增长并不能解决医疗卫生事业所面临的所有问题。当前面临的一个

第五章　护理机器人的伦理考量

The last chapter of the *Chinese Prospects for the Standardization of Robot Ethics 2019* focuses on healthcare robots and analyzes in depth the ethical challenges from the use of robots in medicine, healthcare, rehabilitation, disability assistance, aged-care and so on. It endeavours to provide preliminary advices on how to deal with these challenges in the design, implementation, and supervision of care robots.

We choose care robots as an exemplary case of robot ethics primarily because our society is in urgent need of their application. About one fifth of the global population has motoric, cognitive, or sensory impediments and is in need of medical and healthcare resources. The acceleration of the aged tendency of the human population will undoubtedly increase this proportion. According to the result of the *World Population Prospects: 2017 Revision* issued by the United Nations, there are about 962 million people aged 60 or over in the world in 2017, comprising 13 percent of the global population and the population of elderly people is growing at a rate of about 3 per cent per year. It is also confirmed by the *13th Five-Year National Plan for Developing Undertakings for the Elderly and Establishing the Elderly Care System* that there will be an estimated 255 million people aged 60 or over in China in 2020, comprising 17.8 percent of the country's population. Meanwhile, there is an increasing trend in the percentage of the health expenditure both of GDP and of government expenditure. According to the data from the World Bank, the average percentage of health expenditure of GDP was 9.9 in 2014. In 2016, public spending on health amounted to 700 billion US dollars, taking

严峻挑战是医护从业人员严重不足,世界卫生组织新近发布的《世界卫生统计 2017》显示,我国每万人平均拥有的专业医护人员仅为 31.5 人,远低于美国的 117.8 人和日本的 130.9 人。我国医护从业人员的数量明显无法和当代社会的老龄化趋势相匹配,相关领域面临人力资源短缺的困难。与此同时,最近十年机器人技术在医疗卫生领域的广泛应用,掀起了一场"机器人革命"。在美国、欧洲、日本和我国的机器人技术发展路线图和发展策略中,医用机器人都占有重要地位。可以预见,在不久的将来,人们对健康的刚性需求和医护专业人员的短缺将推动护理机器人井喷式发展。护理机器人因此成为医用机器人的一个重要类别。

其次,护理机器人的出现和广泛使用将会带来深刻的社会和伦理影响。护理机器人的应用,和其他医用机器人一样关系到两个核心概念——健康与关怀。医疗康复手段和照料护理的目的在于帮助人们恢复健康,而护理的价值则首先体现在对病人、残疾人、老人等需要照料和陪护的人群提供必要的关怀。世界卫生组织《组织法》在序言中明确指出,我们今天已经意识到健康不仅仅意味着正常的身体机能,而且是"体格、精神与社会之完全健康状态"。与此同时,对于什么是健康状态,尤其是心理健康的界定,不同的文化和共同体有着不同的理解和实践。医疗语境中的关怀也是如此,它首先是对他人的需求的回应,尤其是对广义的健康需求的回应。[41] 需求有不同的层面,从最基本的生理健康需求、情感需求到个人隐私、人格尊严的保障。不同的文化传统对于什么样的需求应当得到满足,

6.2 percent of GDP. The health market of China is large but still has plenty of space for growth. However, the growth of financial investment cannot solve all the problems related to healthcare. A severe challenge facing China is the lack of a health workforce. According to the World Health Statistics recently released by the World Health Organization (WHO), there are only 31.5 health workers per 10, 000 population in China, much less than 117. 8 in the United States of America and 130.9 in Japan. It is evident that the number of health workers in China cannot cope with the aging trend of the contemporary society. At the same time, the wide application of robotic technology in medicine and healthcare already indicates that a robot revolution is coming. Medical robots play a significant role in the robotics roadmaps of America, Europe, Japan as well as in China. It is not difficult to predict that in the near future there will be an upsurge in the number of care robots because of people's steady demands for health and the lack of health workforce. It is also for this reason that care robots have become one of most important sort of medical robots.

Second, the emergency and wide use of care robots have far-reaching social and ethical implications. As with other sorts of medical robots, the application of care robots is primarily concerned with two core concepts: health and care. The aim of healthcare is to restore health and to offer necessary care to patients, the disabled, elderly people, and so on. It is explicitly mentioned in the 1948 constitution of the World Health Organization that health means not only the level of functional and metabolic efficiency of a living organism, but "a state of complete physical, mental, and social well-being". However, what constitutes a healthy state, in particular the definition of mental health, varies from culture to culture. The meaning of care in a medical context is also diversified in different social communities. The care in question is more than a response to the needs of others, especially the need for health. [41] A person's needs have different levels: from the basic needs of physical health, emotional intimacy to the needs of personal privacy and human dignity. Different cultures have different

什么样的需求具有优先性也有不同的理解。此外,在医疗护理实践中,我们需要面对的是病患、残疾人、老人、幼儿等具有特殊需求的人群。他们往往在情感上对于医疗护理技术尤其是机器人技术的使用高度敏感,但在认知上又容易产生一定的误判。在一定意义上说,他们是更容易受新技术影响也更容易受其伤害的人群,有关机器人的伦理考量理应将他们的权利和价值放到突出的位置。护理机器人因为关系到这些特殊人群和机器人的日常亲密互动,其伦理后果亟待清楚界定和应对。

5.1 护理机器人的界定

依照国际标准化组织(ISO)的界定,机器人根据其用途区分为工业机器人和服务机器人两个基本类别,其中,服务机器人是指在工业自动化应用之外为人类或设备完成有用任务的机器人。服务机器人通常又根据其应用领域和服务对象进一步区分为个人机器人和专业机器人(ISO 8373:2012)。

护理机器人属于服务机器人的范畴,但它的划分标准有所不同,它首先依据的是机器人所服务的主要目的,即保持或改进人类的健康状态,它是一类特殊的医用机器人。因此,护理机器人涵盖一切为人类的疾病、生理损伤、精神伤害及其他生理和心理障碍的康复和护理提供服务的机器人。学界通常根据其使用的情景、发挥的功能及使用者类别来提供一个具有一定解释灵活性的定义:"护理机器人可以定义为在护理实践中使用的,用来满足任何护理需求

answers to what sorts of need should be satisfied and which one has priority over the others. In addition to this cultural diversity, healthcare is significant from an ethical point of view also because it is given to patients, the disabled, and elderly people, who are susceptible to the emotional influences of medical and healthcare practices. These people are also liable to make poor judgements with respect to the genuine nature of high technology. To some extent, they are most vulnerable to the misuse of advanced technology. For this reason, an ethical approach to robotics should place the rights and values of these vulnerable people in a preeminent position. Accordingly, we single out care robots to highlight their possible ethical effects upon those who are in need of a specific sort of care and might form intimate relations with care robots in their everyday life.

5.1　The Definition of Care Robots

According to the definition of International Standard Organization (ISO), robots are distinguished according to their intended tasks into two basic types: industry and service robots. A service robot is "a robot that performs useful tasks for humans or equipment excluding industrial automation application". Service robots are normally further classified by application areas into personal and professional service robots. (ISO 8373:2012)

Care robots are also categorized as service robots, but according to different criteria. What is most significant for their classification is their main purpose: to maintain or improve the health state of human beings. In light of this, care robots are primarily a special sort of medical robot and include all robots that can provide services for the rehabilitation and care of diseases, physical and mental damage and other physiological and psychological impediments. According to current academic discourse, a care robot is often classified in a flexible way according to its context of implementation, its function, and its users: "care robots may be defined as any robot used in a care practice to meet

的机器人，它由护理人或被护理人或者二者共同使用，被用于医院、养老院、临终关怀中心或家庭之类的场合。"[42] 由于健康定义和护理需求的多层次性，护理机器人跨越了个人和专业的区分，既包括家庭中个人使用的助老、助残机器人和具有陪伴和照护功能的机器宠物和仿真机器人等，也包括在专业医疗卫生机构使用的各种用来协助护士完成其日常工作，例如洗浴、喂食、挪移等体力工作的护理机器人。

"护理机器人"这一范畴划分，遵循了世界卫生组织所倡导的健康的宽泛定义，它跨越了传统的医疗和非医疗机构的划分，强调不仅需要通过专业的治疗和康复措施来去除疾病和伤害，而且需要在个人的日常生活中得到恰当的维护。而陪护或者护理所体现的对脆弱人群的关怀，不限于医院、养老所、康复中心等专业机构所提供的一系列专业护理活动，同时还包括所有为了满足老弱病残合理需求的实践，它是一个在包括家庭在内的不同情景中不断延续的完整进程。[43] 这种对健康和护理的整体论理解反映了当代医疗伦理学和护理伦理学的基本价值取向，与我们所推重的繁荣共生原则共融，突出人的身体及其健康的独立价值，以及对整个生活共同体的良性秩序的关注，它也应当成为我们对护理机器人进行伦理反思的出发点。

5.2　护理机器人的一般伦理原则

中国机器人伦理标准化体系包含五个具有国际共识的基本伦

care needs, used by either or both the care-provider or the care-receiver directly, and used in a care context like the hospital, nursing home, hospice or home setting."[42] Due to the multi-layered nature of the definition of health and needs for care, the classification of care robots crosses the distinction of personal and professional robots to include elderly and handicap assistance robots, pet robots and humanoid robots that are designed mainly for personal or domestic use on the one hand, and various sorts of professional service robots that help nurses complete their everyday tasks, such as bathing, feeding, and lifting robots on the other hand.

The classification of care robots follows the WHO's broad definition of health, which insists that the maintenance and improvement of health transcends the traditional division of medical and non-medical institutions: people not only need professional treatment and rehabilitation to cure diseases, but also appropriate care in their daily life. In the same vein, health care for the vulnerable people also goes beyond hospitals, hospice, nursing homes and other professional caring institutions, and involves all domestic practices to satisfy the reasonable needs of the elderly, the disabled and patients, which constitutes a continuous process extended in various context settings, both medical and non-medical.[43] As the basic orientation of contemporary medical ethics and care ethics, the holistic conception of health and care emphasizes the intrinsic value of human body and its health, promote a well-ordered community and agrees with the principle of prioritizing shared flourishing. Such conception of health and care should be adopted as the starting point of our ethical reflections on care robots.

5.2 General Ethical Principles for Care Robots

The Chinese Robot Ethics contains five fundamental ethical objectives that

理目标：人的尊严与人权、责任、透明度、防范滥用和繁荣共生。护理机器人的设计、使用和监管无疑应当为实现这些基本目标服务，从而最大限度地保障我们共同接受的基本价值以及包括机器人在内的整个生活共同体和共生系统最大限度的整体善。与此同时，医疗卫生事业由于直接作用于我们的身体、心灵以及社会交往能力，尤其是针对老年人、孩童、病患、残疾人士等易受外界影响的人群，因此不仅上述基本目标需要细化，而且需要引入专门针对医疗卫生和护理实践的伦理规范。护理机器人的发展和应用也应遵循这一要求。

世界卫生组织2007年发布的《医疗保健，以人为本——协调身心，协调人与系统》一文中，以患者为中心进一步明确了如下基本价值以及由此派生出的根本原则：

1）可及性：即保障人们在需要时可以支配恰当的医疗资源。这进一步明确了人权这一伦理目标中人们平等享有医疗资源的基本权利。

2）安全性：包括在药物、医疗设备、医疗系统、医疗人员操作等环节中保障患者的安全；同时保障服务的透明度和明确相关的责任范围，避免相关服务被滥用。这是对透明度和防范滥用目标的进一步细化。

3）质量：它包括保健的内容（即循证指导方针与实际建议相结合）和过程（即效率、适时与协调），确保人们享有高效优质的医疗卫生服务。

4）可负担性：公共卫生服务的费用应该是政府和用户可以负担

have international consensus: human dignity and rights, accountability, transparency, awareness of misuse, and prioritizing shared flourishing. The design, implementation, and supervision of care robots should put into effect these basic principles to maximize the fundamental values we all cherish. In the meanwhile, medical and healthcare practices directly affect our body, mind and social communication, with a special reference to patients, the disabled, the elderly, children and other vulnerable people. Therefore, in addition to concretizing the general principles mentioned above, we also need to introduce ethical norms that are specially designed for healthcare practices to regulate the development of care robots.

In 2007, the World Health Organization issued a document entitled *People at the Centre of Health Care: Harmonizing Mind and Body, People and Systems,* which clarifies the fundamental values and principles of patient-centered healthcare:

1) Access: a health system should secure the accessibility to appropriate health resources when needed. This principle makes explicit the basic right of people to have equal access to health care.

2) Safety: it includes the safety of a patient in medicaments, medical devices, and a health system. This principle also aims to guarantee the transparency of health service and its accountability and to avoid the misuse of health service.

3) Quality: it aims to provide health service of high efficiency and quality, with regard to both the content (i.e. adherence to evidence-based guidelines and practice recommendations) and the process (i.e. efficiency and timeliness and coordination) of care.

4) Affordability: the cost of public health service should be affordable for the government and the people, which safeguards the equity and justice in

的，维护社会资源分配的公平和正义，避免与个人生存健康相关的人的基本权利受到威胁。

5）满意度：世界卫生组织创造了"反应性"这一概念来说明医疗系统如何调动各种资源来针对人们的相关需求做出反应，从而帮助我们具体评估包含着患者主观经验的满意度。它包含七个基本要素，进一步细化了医疗系统应当遵循的基本规范，保障我们所追求的基本伦理目标能够落实到具体的医疗实践中：

a）对个人尊严的尊重；

b）机密性：患者对何人有权访问其个人医疗记录的决定权；

c）患者享有参与个人健康选择的自主权；

d）及时关注：对急救的迅速关注以及非急救的合理等待时间；

e）优质的福利设施，例如清洁、空间与医院食物；

f）接受医疗的患者可以获得社会支持网络，例如家庭与朋友；

g）患者享有选择由哪个个人或哪家组织来提供医疗的自由。

世界卫生组织的上述原则全方位地考察了患者或医疗保健服务消费者的基本权利，护理机器人的设计、研发、应用、宣传和监管都需要充分考虑这些基本规范，确保被照护者的基本价值得到维护。

5.3 护理机器人的不同存在方式及其伦理后果

要将上述抽象的伦理规范转换为护理机器人的设计和研发中切实可行的有效约束，尤其是要推动伦理价值在机器人技术运用中的

distributing social resources and protects the rights of a person's existence and health.

5) Satisfaction: WHO developed the concept of "responsiveness" to show how a health system should employ various sorts of resources to make appropriate responses to people's health needs. It is a useful instrument to evaluate the subjective satisfaction of a patient with the health system in a concrete manner. It has seven distinct elements that stipulate the ethical norms for a health system.

a) respect for the dignity of the person.

b) confidentiality, or the right to determine who has access to one's personal health records.

c) autonomy to participate in choices about one's own health.

d) prompt attention: immediate attention in emergencies and reasonable waiting.

e) amenities of adequate quality, such as cleanliness, space and hospital food.

f) access to social support networks, such as family and friends, for people receiving care.

g) choice of provider, or freedom to select which individual or organization delivers one's care.

These detailed principles offered by WHO give a comprehensive survey of the basic rights of a patient or a consumer of health service, which should be carefully considered and defended in the design, development, implementation, advertising, and supervision of care robots.

5.3　Different Modes of Care Robots and Their Ethical Implications

To transfer the abstract ethical norms listed above into practicable policies that can be implemented in the design and development of care robots, it is neces-

实现，依照此前论述的中国机器人伦理标准化体系的五步骤实施方法，我们认为有必要从不同视角来审视护理机器人存在和发生作用的不同方式。它可以帮助我们更系统地展示护理机器人在人类医疗卫生实践的不同层面上产生的伦理后果以及相应的应对措施。

1）护理机器人首先是物理存在：它是被人类有意识地制造、具有特定功能、可以协助或者取代人类完成特定任务的工具。和其他人造工具或装备一样，护理机器人在这个层面的伦理问题主要涉及操作安全性和环境友好性，它关系到机器人伦理风险评估列表中的个人、环境与共生系统两个层面由机器人的物理活动所带来的伦理危害。例如用于脑瘫儿童康复的机器人应当装备防止儿童使用的安全设置；植入人体的材料要经过生物相容性评估等安全性测试，其他相关材料也应对环境友好；护理机器人的程序设计要尽可能避免出现故障；家用护理机器人的操作应当便于用户理解，确保老年人和残障人士等在没有专业人士的指导下也能够便捷安全地使用机器人；护理机器人研发和宣传机构应当充分说明机器人可能的安全隐患和其他潜在危害，确保机器人服务对象享有充分知情同意权等等。在物理存在这个层面上，上述与安全性相关的伦理问题大多可以通过引入或建立相关的国际国内行业设计和生产标准、制定相应的政策法规来予以解决，保障护理机器人使用者的人身安全、维护环境可持续性发展原则以及生态多样性原则。此外，机器人的程序设计和操作机制也应当保证充分的透明度和可理解性，以便在机器人故障造成人身伤害或财产

sary to adopt the five-step process methodology in the Chinese system for the standardization of robot ethics. We thus should examine the different modes of care robots' functioning, which will secure a systematic approach to their possible ethical effects and necessary responses to their ethical challenges on different levels of healthcare practices.

1) A care robot is above all a physical being: it is a tool or instrument that is deliberately designed to assist or replace human beings to perform certain intended tasks. Like other artificial tools or equipment, a care robot's ethical problems are mainly concerned with the safety of operation and the friendliness to environment. These problems represent ethical risks which robots may produce with respect to individuals and environment and ecosystem. For instance, a robot designed for the rehabilitation of cerebral palsy children should be equipped with devices that can easily terminate its operations in unexpected situations; the materials implanted in the human body should pass safety tests such as the evaluation of biological compatibility, while other materials should be friendly to environment as well; the program design of care robots should make every effort to avoid possible breakdown, which might be harmful for patients; domestic care robots should be easy to operate so that the elderly people and the disabled can safely use robots without the guidance of a professional health worker; the institutions in charge of designing and promoting care robots should carefully publicize their safety hazards and other potential harms, in order to guarantee the informed consent of the service receivers and so on. On the physical level, most of the ethical problems related to safety can be solved by introducing professional standards of design and product, or issuing specific policies and laws to protect the safety of the users of care robot and the principle of environmental sustainability. Furthermore, the program design and operation mechanism should also secure

损失时可以清晰地追溯和判定责任来源。除了首要的安全性问题外，护理机器人作为一个物理存在，在其外观设计和操作使用上应杜绝任何形式的歧视（年龄、种族、性别、宗教信仰歧视等），应能满足不同文化传统的多样性需求。

2）护理机器人作为医疗卫生系统的一个构成要素，它也是一种社会资源。在引入机器人来提高我们的工作效率的同时，我们必须考虑它是否会威胁到社会的公平和正义。对照我们的伦理风险评估列表中社群与社会、经济与商业这两个层面的伦理隐患，这里至少有两个方面的问题需要考虑，一是护理机器人作为稀缺医疗资源的分配问题。护理机器人对于保障病患和残障人士的健康需求和基本生活质量意义重大，它们并不单纯是卫生保健消费品，它们中的一部分已经开始成为公共医疗资源的一个重要组成部分，例如在日本医院中投入使用的具有挪移功能的机器人。公共卫生保障系统需要保证资源的合理配置，使有需要的病患能够得到并且能够支付相应的机器人服务，而不是进一步扩大医疗服务上已经存在的贫富差距。前文提及的可及性和可支付性是这个层面需要捍卫的首要伦理原则。例如护理机器人的运行依赖程序的更新，这些程序更新所产生的费用不仅要考虑市场的需要，也需要考虑患者的承受能力，否则将威胁到人人健康的基本价值诉求。护理机器人的适用对象往往是老、幼、病、残这样的弱势群体和敏感群体，相关决策尤其需要谨慎。另一个需要考虑的方面是：机器人的引入在减轻医护人员工作负担的同时，也可能对相关行业的就业造成冲击。我们需要对大规模引

the transparency and intelligibility so that one can easily trace and judge the origin of responsibility when the breakdown of a care robot brings about personal harm and property damage. In addition to the worries about physical safety, the appearance and the operation of a care robot should avoid any form of discrimination (age, race, gender, religious beliefs, and so on) and satisfy various needs of different cultural traditions.

2) As a component of a health system, care robots are also a sort of social resource. When we introduce care robots to increase the efficiency of health care, we must also consider if they will become a threat to social equity and justice. With respect to "community and society" and "economics and business" in the table above, at least two ethical issues should be addressed on this level: the first is the distribution of care robots as a rare health resource at the moment. Care robots can bring significant improvement of life quality for patients and the disabled people. They are not only healthcare consumable products. Some of them have become an important components of public health resources, such as the robots for lifting in Japanese hospitals. A public health system has the duty to ensure reasonable allocation of resources so that a patient in need of the service of a care robot should be able to have access and to afford it. The distribution of care robots should not enlarge the existing gap between the rich and the poor with regard to health service. Accessibility and affordability are the most important ethical principles on this level. For example, the good functioning of care robots relies on the updatable programs. However, the relevant cost should not only consider the market's needs, but also the affordability of patients; otherwise it would threaten the fundamental value of health for everyone; the receivers of care service are primarily vulnerable people like the elderly, the disabled, children and patients. Therefore, one should be very cautious in making decisions in this regard. The second ethical issue is the impact of introducing of care robots on job markets. The government should

入护理机器人，尤其是护理机器人的经济和社会后果进行评估。一种可能的应对方案是要明确护理机器人所充当的助理角色及其投放领域和规模，尽可能避免用护理机器人完全替代医护人员，而是要求在医护人员监管下使用机器人，这同时也有助于降低机器人在上述操作层面存在的风险。

3）机器人作为数据的载体存在，它需要获取、存储、传输和利用数据。护理机器人直接接触与个人身体、心理、社交等相关的数据，而且可以出现在家庭这样的私人场所，它会接触和搜集大量的个人身份信息，由此产生的信息安全和隐私问题尤为严峻，参照我们的伦理风险评估列表，这是个人与个体层面需要面对的最严重的伦理危害。机密性、隐私权和知情同意权是这个层面首先需要捍卫的基本原则，而这些基本权利也是人的自主权和人格尊严得以捍卫的重要保障。具体而言，护理机器人的生产厂商和相关医疗护理机构需要向患者充分说明他们搜集的是哪一类信息，说明使用相关信息的必要性，充分保障用户的知情同意权。物联网和虚拟技术的出现使得人们的个人信息应用愈发广泛，个人数码人格和身份的形成指日可待，其中与健康相关的数据记录至关重要。我们有必要在护理机器人的设计和使用中充分保障个人对其数码人格和身份的控制和支配，我们也应推动相应的立法工作，来监管包括政府在内的相关机构对个人数据的使用，明确在什么情景下个人健康信息可以被合法获取和使用。

make careful evaluations of the economic and social implications of introducing a large number of care robots. A possible solution is to explicitly define the assistant role and the amount of care robots, and not to allow care robots to replace health workers. It can be required that care robots be always used under the guidance and supervision of a professional caregiver. Incidentally, this requirement can help reduce the safety hazards on the physical level.

3) Robots as bearers of digital data need to obtain, save, transmit and employ the digital data. Care robots have direct contact with personal physical, psychological, and social information and can appear in private locations such as a person's house. They can collect an enormous amount of personal identity information that creates a severe challenge to the information safety and the protection of one's privacy. Privacy constitutes one of the most crucial ethical concerns with respect to person and individual. Confidentiality, privacy and informed consent are the basic principles that should be defended on this level, while these rights also constitute the basis of a person's autonomy and dignity. The producers of care robots and correspondent health institutions should sufficiently explain to patients the necessity of collecting their information and what sort of information will be collected so that the patients can make a decision and choice based upon informed consent. With the appearance of the internet of things and the wider application of virtual technology, the formation of a person's digital persona and identity is coming in the near future, of which digital records related to one's health state comprise a significant part. In the design and use of care robots, a person should obtain absolute control over his or her digital personas and identities. We should also accelerate the process of legislation to supervise the institutional use of personal data, including governmental use as well. It should be clearly defined under what situations one's health information can be legally obtained and used.

4）机器人也是具有一定自主性的人工智能体。与其他物理存在和社会资源非常不同，机器人接受和处理数据的能力使它不再仅仅是人类社会活动完全被动的参与者。机器人拥有三个核心组件：传感器可以用来监控和接受外部（包括宿主）的刺激或信息，处理器则决定如何对之做出回应，效应器或驱动器则在一定意义上执行这些决策，对环境发生作用[44]。机器人可以被视为"处身于世界之中能够感觉、思考和行动的机器"。[45] 具有一定学习能力的机器人尤其如此。当它们不仅能够通过状态的变化来对刺激做出反应，而且能在没有刺激的情况下改变自己的状态，甚至有可能通过对自己经验的学习和分析，改变自己与环境互动的转换规则，这时它就具备了所谓"人工智能体"的三个重要特征：互动性、自主性与适应性。[46] 在伦理实践中如何看待这种能够接受和处理与若干情境相关的伦理信息，并且能够做出相应决策和行动的机器人，仍然是学界激烈争辩的话题。[47] 目前无可争议的是，这些人工智能体或人工能动者并不是完全的道德能动者，它们不能为自己的行动承担道德责任。对它们来说，我们对它们的行为所做出的道德反应，例如赞扬或谴责并没有任何实质的意义。同样没有太多争议的是，在机器人与人构成的伦理网络中，机器人即使不应被看作独立的行动者，也应被看作构成性的道德要素。它可以根据所获得的信息改变环境，并且通过与人互动，产生在道德上或好或坏的后果。简单来说，机器人作为伦理网络的参与者，它的设计和使用本身就不再是价值中立的，而是承载着一定的道德价值。

4) Robots are also artificial agents that have a certain amount of autonomy. Unlike other physical beings and social resources, robots are not purely passive participants on human being's social life thanks to their ability to accept and process data. A robot has three core components: sensors for monitoring and receiving external (including the host) stimulations or information, processors for determining how to respond to them, and effector or actuator for carrying out these decisions and to affect the environment.[44] In light of this, robots can be taken as machines that can perceive, think and act in the world.[45] This is especially true with the robots that have a certain capability of learning. If robots can not only respond to stimulations by making reactions, but also change their own states without being stimulated, and even change the rules of interaction with the environment by learning and analyzing, then they have three essential characteristics of artificial agents: interactivity, autonomy, and adaptability.[46] It is still a highly controversial issue how to treat the robots that can receive, process and deal with contexts related ethical information and make certain decisions to act.[47] What is uncontroversial is that these artificial agents or agents are not full moral agents and therefore cannot take full moral responsibility for their actions. Our moral attitudes or reactions to their actions, such as praise or blame, make no sense for them. On the other hand, where there is little controversy is that robots should be taken as moral factors, even though not independent agents in the ethical networks of robots and human beings. Robots can change the environment by their information about the environment and bring about morally good or bad consequences by their interactions with human beings. In short, the design and use of robots as participants in the ethical network cannot be taken as morally neutral, but is necessarily value loaded.

以这种方式存在的护理机器人，会给我们带来更加严峻的伦理挑战。例如，我们能否为护理机器人预先嵌入伦理代码？我们是否允许护理机器人具有通过深度学习来形成自主决策的能力？当机器人以其设计者和制造者未能预知的方式造成伤害时，我们如何确定其责任归属？此外，在我们的伦理网络中，情感是行动者之间进行交流的重要纽带，我们是否应当鼓励人和机器人之间建立情感纽带？因为机器人作为智能体的状态仍然含混并富有争议，机器人和人的互动仍然处于萌芽阶段，我们还需要以跨学科方式对这些伦理问题进行更深入的研究，重新评估作为人工智能体的机器人在个人与个体和社群与社会层面带来的伦理挑战。

世界卫生组织所倡导的回应性概念，突出患者或医疗服务使用者的主观满意度。因此，在探讨护理机器人作为一个道德因素所带来的伦理问题时，我们要充分尊重接受护理的用户的知情权和参与权。我们倡议建立相关的研究中心，创立高效的交流平台，邀请具有不同学科背景的学者、研发者、医疗卫生从业人员、护理机器人的潜在用户代表进行更深入的交流和讨论，在充分了解相关伦理问题的复杂性基础上提出相应的应对措施。

5.4 护理机器人伦理问题例示：智能假肢和机器宠物

在充分了解护理机器人相关的基本伦理规范和护理机器人在不同的存在层面需要面对的伦理问题的基础上，我们以智能假肢和机器宠物为例来进一步展示在特殊类别的机器人使用上有可能遭遇的

On this level, care robots bring more serious ethical challenges to human beings. For instance, can we embed in advance moral codes in a care robot? Shall we permit it to have the power of deep learning to form a capacity to make autonomous decisions? When a robot brings about harms unexpected by its designers and producers, how should we ascribe the responsibility? Moreover, emotions are a significant link between human agents in our ethical network. Shall we encourage the emotional interactions between human beings and robots? Since the status of a robot as an agent is still ambiguous and controversial, and the interactions between human beings and robots are still at a primitive stage, we need more in-depth and interdisciplinary inquiry into the ethical nature of these issues and evaluation of ethical challenges with respect to both individual and community.

The notion of responsiveness promoted by WHO underlies the subjective satisfaction of patients with health service. Therefore, we should respect the right of a service receiver to know and to participate in the design of care robots. We urge the society to establish research centers and efficient platforms, and to invite scholars, developers, health workers, potential consumers of care robots to engage themselves in more profound communications and discussions, and to make appropriate measures to cope with the ethical challenges.

5.4 Artificial Prostheses and Robot Pets

We choose artificial prostheses and robot pets as examples to show more complex ethical challenges and correspondent measures in some special sorts of robots, which go beyond the basic moral norms and ethical problems on different levels of the existence of a care robot.

更为复杂的伦理挑战和应对措施。

智能假肢是一种功能替代性的助残机器人,它采用了机器人技术来为肢体残缺人士提供行动能力和操作能力。例如某智能动力小腿假肢,它是一种可穿戴设备,通过传感器来精确采集穿戴者的神经信号,快速做出响应;通过芯片计算结果来适时判断穿戴者所处的地理环境,推测使用者的动作意图,直接做出决策,对助残肢体下达动作指令。在穿戴者行走的关键阶段,还能提供主动力矩,成功实现对多种地形的主动适应。

作为一种可穿戴设备,假肢机械结构的本体与穿戴者以及外部环境存在着很强的物理交互。因此在复杂的环境中保持稳定的性能和良好的安全性无疑是智能假肢设计的重要诉求;它的故障很可能会对残疾人士造成十分严重的伤害。这里值得一提的是智能假肢作为物理存在的方式与其他护理机器人有所不同。它和人体这个特殊存在有着更为紧密的关联。一些智能假肢的使用者声称他们感觉这些具有自然肢体运动能力的假肢成了他们身体的一部分。有些智能假肢也能通过各种各样的感受器向穿戴者进行感知反馈,此时损害假肢也在同时损害穿戴者的自然身体。而在我们的伦理实践中,身体因为与自我的亲密关联具有特殊的价值。例如在司法领域,对身体的伤害和对其他物体的损害属于不同的范畴。如果智能假肢被普遍接受为穿戴者身体的一部分,那我们相关的司法和伦理实践都要发生改变。

此外,目前有的智能假肢已经展示出自然肢体所不具备的能力,例如已经投入使用的某种模块式假肢,它的穿戴者可以轻易地独立

Artificial prostheses are assistive robots that are used as substitutes for the functions of missing bodily limbs. They adopt robotic technology to provide motoric and operational capabilities to the disabled people. For example, an intelligent prosthesis for the missing lower limb can collect neural signals of its wearer by sensors and make immediate response. It can make real time judgments about the geographical situations of the wearer and infer his or her intentions to move and make immediate decisions and ordering the assistive limb to move. At the crucial stages of walking, it can offer active momentum and actively adapt to various forms of terrain.

As a wearable equipment, the mechanical structure of the prosthesis has strong physical interactions with external environment. What matters for the design of an artificial prosthesis is the stable functioning and high safety in complex situations: its failure can bring severe or even mortal harm to the disabled. It deserves notice that the physical being of artificial prostheses is distinct from other care robots: it has an intimate relation with the human body. Some users of artificial prostheses have claimed that they feel these mechanical devices have already became a part of their bodies. Some of these artificial devices can return sensory reactions to the wearers. In this case, doing damage to the devices is also doing damage to the natural human body. However, the body plays a unique role in our ethical practices because its intimate relation to the self. For example, the harm of the body of another person and the harm of that person's properties belong to two different categories. If artificial prostheses can be universally taken as a part of the bodies of their wearers, our legal and ethical practices should change accordingly.

Moreover, some artificial prostheses also show some capacities that cannot be accomplished by natural limbs. For instance, a sort of modular prosthetic limb can enable its user to bend each of its "fingers" with the others remaining

弯曲假肢的任何一个手指，而其他手指保持不动，这是任何人手都无法自然完成的动作。我们不难设想，在不远的将来，智能假肢不仅能恢复缺失的自然肢体的机能，甚至能在某些方面起到增强穿戴者能力的作用。在那样的条件下，我们是否允许健康人用智能假肢替换健康的肢体？我们是否允许人们随意地处置自己的身体，就像处置自己所拥有的财产一样。与目前流行的以人类中心主义为基础的机器人伦理主张不同，中国机器人伦理标准化体系突出繁荣共生的原则，强调身体作为自然物有着不可取代的内在价值，它是我们的人格及其尊严的重要构成部分，它为我们提供了充分的理由限制我们随意地使用智能假肢替换我们正常的身体部件，可以有效地防范由此带来的伦理危害。

与智能假肢不同，智能机器宠物所带来的伦理问题更集中在它与使用者的互动上。需要说明的是，智能机器宠物本身不一定以健康护理为目的，它很可能只是作为满足个人娱乐需求的玩具机器人而进入市场。然而实践表明这一类机器人往往因为其出色的社交功能而具有陪伴、护理甚至辅助心理治疗的能力。我们有充分的理由将它划入护理机器人的范畴，因为它在保持和促进病患的心理健康上发挥着不可替代的作用，而且它是未来具有更强自主活动能力和互动能力的仿真护理机器人的先导。例如某种机器狗，它的最初版本于 2005 年投入市场，能够行走，感知距离、声音和压力，能够与人进行简单的互动。而 2018 年推出的新版本则能够识别和执行简单的英文指令，例如握手和坐下，动作也更为灵活，而且能够感受主

unmoved, an action that cannot be realized on the human body. It is not difficult for us to imagine that in the near future artificial devices can not only recover the lost functions of a natural limb, but even efficiently enhance the wearer's capacities. In that case, shall we allow a healthy person to substitute his natural limbs with artificial ones? Shall we allow a person to deal with his own body, just as a part of his own properties? In contrast to the prevalent anthropocentrism, the Chinese system for the standardization of robot ethics stresses the principle of shared flourishing and the intrinsic value of human body which as natural organism is the carrier of personality and human dignity at the same time. The irreplaceable natural value of human body provides us with a perspective from which the beneficial use of artificial prostheses can be differentiated from the harmful one.

Unlike intelligent prostheses, the ethical problems created by robot pets are concentrated on the interaction between them and their users. It should be emphasized that intelligent pets are not necessarily designed for the sake of health care. They may be introduced into markets as entertainment robots or toys. However, medical practices show this sort of robots has unusual capabilities of accompanying, giving care to, and assisting the psychological therapy of a patient by its excellent social functions. We have good reasons to take it as a special sort of care robots because it plays an unreplaceable role in maintaining and improving the mental health of a patient and is a forerunner of more autonomous and more interactive humanoid robots in the future. For example, a robot dog that was firstly introduced into the market in 2005 has the capacities of walking, perceiving distances, sounds and pressures, and has very simple interactions with human beings. The new version of this robot dog issued in 2018 can better recognize and carry out simple commands in English, such as shaking hands and sitting down. It can

人的触摸，识别主人的声音和面孔，甚至还能随着时间的推移通过与主人的互动形成独特的个性。有研究表明该机器狗的使用可以改善老年痴呆症患者的活动和社会行为，而自闭症儿童也愿意同机器狗进行口头互动和交互互动。[48]

这类机器宠物在医护领域广受欢迎的同时，也引发了研究者的担忧。最近的问卷调查表明，多数回答者表达了对自闭症患者对社交机器人产生情感依赖的担忧，他们希望患者和机器人的互动得到治疗师的指导，同时也希望机器人被远程控制，而不是完全自动化。[49] 有伦理学家也认为向患者提供这样的机器人陪护，是在给他们提供虚伪的朋友。[50] 这里的忧虑涉及伦理风险评估中个人与个体、社群与社会两个领域，呈现为三个不同的层面：

1）情感依赖。尤其是患者以拟人化的方式对待机器人，从而将情感寄托在不具有真正意识和生命的事物上。但是，如何评估这样的隐患以及人们的忧虑取决于我们如何看待机器人在伦理交往中的地位，它是否能被视为一个独立的能动者。它也取决于我们如何理解情感交往，真正的情感交往是否只能发生在两个具有意识的生命体之间。而我们之前在关于机器人作为自主道德智能体的讨论中已经看到，类似这样的问题目前都是有争议的。

2）欺骗问题。机器宠物在与人的交往中表现为与它实际的存在方式不同，从而给用户尤其是认知上存在一定障碍的用户带来了不必要的混淆；这种混淆很可能是有害的，它让机器宠物的主人误以为能够从机器宠物那里得到真正的情感回应，然而实际上所获得的

also perceive the caress of its owner and recognize its owner's face and voice. More importantly, it can develop a specific character by its interactions with the owner in the longer term. It has been confirmed by some recent research that the use of robot dogs can improve the motoric and social activities of Alzheimer's patients, and that autistic children are willing to have oral and mutual interactions with these dogs.[48]

This sort of robot pets also gave rise to worries among researches. Some recent questionnaires show the worries about the emotional reliance of autistic patients upon social robots. Most people hope to bring the interactions of patients and robots under the guidelines of therapists. They also believe it is better to employ remote controlled robots rather than fully automatic ones.[49] There are also some philosophical ethicists arguing that this is equal to offering counterfeit friends to patients.[50] These ethical risks which concern both "person and individual" and "community and society" manifest themselves on three different levels:

1) Emotional dependency. This is especially the case when a patient anthropomorphize robots and entrust their emotions to things that do not have genuine consciousness and life. However, whether this is really an ethical problem depends on how we evaluate the role of robots in ethical communications, and on whether they can be taken as independent agents. It also depends how we understand affective interactions, and whether genuine emotional exchanges only happen between two conscious living beings. It is already obvious in our discussion on the status of robots as autonomous moral agents that these issues are still hotly debated.

2) Deception. Robot pets appear to be other than what they really are in their communications with human beings, which brings about unnecessary cognitive confusion to their users. This confusion can be harmful because

只是预先设定的机械回应。

3）人机交往对人际交往的影响。自闭症患者非常有可能在人机交往中获得更多回报，以致进一步远离人与人之间的复杂交往，反而加重病情。而且人机交往还会破坏社会生态的多样性，使机器宠物的使用者接受单一类型的机器回应，从而进一步丧失社会交往能力。

当然，除了这三个直接的伦理忧虑之外，机器宠物使用者的隐私、自主性等基本价值也会受到威胁。我们的伦理风险列表已经对大多数风险进行了充分的考量，并给出了有效的应对方式。例如，我们可以通过避免不必要的拟人化设计、增强对相关心理成瘾问题严重性的意识等措施来防范患者对机器宠物的不当依赖；我们应当确保机器宠物的设计理念和意图能够清晰准确地传达给用户，并且通过职业医疗护理人员的介入和指导，保证患者能够正确理解机器人给出的与人相似的情感回应，避免患者受到欺骗或错误理解社会交往准则；此外，我们反复强调的繁荣共生原则要求我们将机器人看作人类生活共同体的一个有机构成部分，并要求我们善待机器宠物，避免滥用甚至虐待机器宠物从而使患者本人的伦理品质也受到伤害。当然，如何应对人机情感交互这类前沿话题，我们无疑还需要更加深入的研究，倾听利益相关各方的声音，从不同的学科立场出发进行评估和衡量。目前而言，我们至少应当坚持，当机器宠物用于医疗领域时，一定要在具有专业资质的治疗师指导下进行；毕竟在机器人尚未成为完全自主的行动者之前，人与人的交往仍然是我们伦理生活最基本的形式。

the owners might wrongly believe that they can obtain genuine emotional reactions while what they get is nothing more than prearranged mechanical responses.

3) The impact of person-machine interaction on person-person interactions. Autistic patients are liable to gain more rewarding experiences from interactions with machines, which keep them from more complex communications between human persons.

In addition to these three ethical worries, the privacy, autonomy and other basic ethical values of the owner of a robot pet can be threatened as well. In the last chapter, we have listed the ethical risks coming along with the deployment of robots and their mitigations. For example, we can avoid an unnecessary anthropization and educate an awareness of addictiveness in order to prevent the improper attachment of patients to robot pets. Further, we should make the idea and intention of robot pet design accessible to user. In the meanwhile, professional therapists may help patients properly understand an appearance of human-like emotional reaction and keep the latter from being deceived and misinterpreting social conventions. Finally, robots must be regarded as a sort of components of human community on the basis of the principle of shared flourishing. Patients are thus obliged to well treat robot pets, avoid their misuse and abuse, and keep from any harm of their own ethical characters. In order to deal with this cutting-edge topic, we need to launch more profound research projects to carefully listen to various voices from different backgrounds. As regards our current situation, it is wiser to insist that when robot pets are used in health practices, they should be placed under the guidance of professional therapists. After all, when robots are not yet full autonomous agents, the most basic form of ethical life is the communications among human persons.

参考文献

1. European Parliament, Report with recommendations to the Commission on Civil Law Rules on Robotics. http://www.europarl.europa.eu/sides/getDoc.do?type=REPORT&mode=XML&reference=A8-2017-0005&language=EN（访问时间 2018 年 4 月 30 日）

2. EGE, *Statement on Artificial Intelligence, Robotics and 'Autonomous' Systems.* https://ec.europa.eu/research/ege/pdf/ege_ai_statement_2018.pdf（访问时间 2018 年 4 月 30 日）

3. The European Commission's High-Level Expert Group on Artificial Intelligence, *Draft Ethics Guidelines for Trustworthy AI,* https://ec.europa.eu/digital-single-market/en/news/draft-ethics-guidelines-trustworthy-ai（访问时间 2018 年 4 月 30 日）

4. J. J. Prinz, *The Emotional Construction of Morals,* Oxford: Oxford University Press, pp. 270-273, 2007.

5. P. Lin, "Introduction to Robot Ethics", in *Robot Ethics: The Ethical and Social Implications of Robotics,* ed. by Patrick Lin, Keith Abney and George A. Bekey, Cambridge: The MIT Press, 2012, pp. 3-15.

6. S. di Robotica and G. Veruggio, *EURON Roboethics Roadmap,* 2006.

7. G. A. Bekey, "Current Trends in Robotics: Technology and Ethics", in *Robot Ethics: The Ethical and Social Implications of Robotics,* ed. by Patrick Lin, Keith Abney and George A. Bekey, Cambridge: The MIT Press, 2012, pp.

References

[1] European Parliament, Report with recommendations to the Commission on Civil Law Rules on Robotics. http://www.europarl.europa.eu/sides/getDoc.do?type=REPORT&mode=XML&reference=A8-2017-0005&language=EN (visited on April 30,2018)

[2] EGE, *Statement on Artificial Intelligence, Robotics and 'Autonomous' Systems.* https://ec.europa.eu/research/ege/pdf/ege_ai_statement_2018.pdf (visited on April 30,2018)

[3] The European Commission's High-Level Expert Group on Artificial Intelligence, *Draft Ethics Guidelines for Trustworthy AI,* https://ec.europa.eu/digital-single-market/en/news/draft-ethics-guidelines-trustworthy-ai (visited on April 30,2018)

[4] J. J. Prinz, *The Emotional Construction of Morals,* Oxford: Oxford University Press, pp. 270-273, 2007.

[5] P. Lin, "Introduction to Robot Ethics", in *Robot Ethics: The Ethical and Social Implications of Robotics,* ed. by Patrick Lin, Keith Abney and George A. Bekey, Cambridge: The MIT Press, 2012, pp. 3-15.

[6] S. di Robotica and G. Veruggio, *EURON Roboethics Roadmap,* 2006.

[7] G. A. Bekey, "Current Trends in Robotics: Technology and Ethics", in Robot *Ethics: The Ethical and Social Implications of Robotics,* ed. by Patrick Lin, Keith Abney and George A. Bekey, Cambridge: The MIT Press, 2012, pp.

17-34.

8 *The Economist,* "Trust me, I'm a robot", 2006. https://www.economist.com/node/7001829（访问时间 2018 年 4 月 30 日）

9 G. A. Bekey, "Current Trends in Robotics: Technology and Ethics", in *Robot Ethics: The Ethical and Social Implications of Robotics*, ed. by Patrick Lin, Keith Abney and George A. Bekey, Cambridge: The MIT Press, 2012, pp. 17-34.

10 G. A. Bekey, P. Lin, K. Abney, "Ethical Implications of Intelligent Robots", in *Neuromorphic and Brain-based Robots*, ed. by Jeffrey L. Krichmar and Hiroaki Wagatsuma, Cambridge: Cambridge University Press, 2011, pp. 323-344.

11 G. A. Bekey, "Current Trends in Robotics: Technology and Ethics", in *Robot Ethics: The Ethical and Social Implications of Robotics*, ed. by Patrick Lin, Keith Abney and George A. Bekey, Cambridge: The MIT Press, 2012, pp. 17-34.

12 转引自 P. Lin, R. Jenkins, K. Abney, *Robot Ethics 2.0*, Oxford: Oxford University Press, 2017。

13 转引自 P. Lin, "Introduction to Robot Ethics", in *Robot Ethics: The Ethical and Social Implications of Robotics*, ed. by Patrick Lin, Keith Abney and George A. Bekey, Cambridge: The MIT Press, 2012, pp. 3-15。

14 IFR, "How Robots Conquer Industry Worldwide", 2017. https://ifr.org/downloads/press/Presentation_PC_27_Sept_2017.pdf（访问时间 2018 年 4 月 30 日）

15 参见 IFR，"Executive Summary World Robotics 2017 Industrial Robots"。https://ifr.org/downloads/press/Executive_Summary_WR_2017_Industrial_Robots.pdf（访问时间 2018 年 4 月 30 日）

16 参见 IFR，"Executive Summary World Robotics 2017 Service Robots"。

17-34.

[8] *The Economist*, "Trust me, I'm a robot", 2006. https://www.economist.com/node/7001829（visited on April 30,2018）

[9] G. A. Bekey, "Current Trends in Robotics: Technology and Ethics", in *Robot Ethics: The Ethical and Social Implications of Robotics,* ed. by Patrick Lin, Keith Abney and George A. Bekey, Cambridge: The MIT Press, 2012, pp. 17-34.

[10] G. A. Bekey, P. Lin, K. Abney, "Ethical Implications of Intelligent Robots," in *Neuromorphic and Brain-based Robots,* ed. by Jeffrey L. Krichmar and Hiroaki Wagatsuma, Cambridge: Cambridge University Press, 2011, pp. 323-344.

[11] G. A. Bekey, "Current Trends in Robotics: Technology and Ethics", in *Robot Ethics: The Ethical and Social Implications of Robotics,* ed. by Patrick Lin, Keith Abney and George A. Bekey, Cambridge: The MIT Press, 2012, pp. 17-34.

[12] Cf. P. Lin, R. Jenkins, K. Abney, *Robot Ethics 2.0,* Oxford: Oxford University Press, 2017.

[13] P. Lin, "Introduction to Robot Ethics," in *Robot Ethics: The Ethical and Social Implications of Robotics,* ed. by Patrick Lin, Keith Abney and George A. Bekey, Cambridge: The MIT Press, 2012, pp. 3-15.

[14] IFR, "How robots conquer industry worldwide", 2017. https://ifr.org/downloads/press/Presentation_PC_27_Sept_2017.pdf（visited on April 30,2018）

[15] IFR, "Executive Summary World Robotics 2017 Industrial Robots" .https://ifr.org/downloads/press/Executive_Summary_WR_2017_Industrial_Robots.pdf（visited on April 30,2018）

[16] IFR, "Executive Summary World Robotics 2017 Service Robots". https://ifr.

https://ifr.org/downloads/press/Executive_Summary_WR_Service_Robots_2017_1.pdf（访问时间 2018 年 4 月 30 日）

17 G. Veruggio and F. Operto, "Roboethics: Social and Ethical Implications of Robotics", in the *Springer Handbook of Robotics*, ed. by Bruno Siciliano and Oussama Khatib, Berlin: Springer, 2008, p. 1504.

18 K. Abney, "Robotics, Ethical Theory, and Metaethics: A Guide for the Perplexed", in *Robot Ethics: The Ethical and Social Implications of Robotics,* ed. by Patrick Lin, Keith Abney and George A. Bekey, Cambridge: The MIT Press, 2012, p. 35.

19 W. Wallach and C. Allen, *Moral Machines: Teaching Robots Right from Wrong*, Oxford: Oxford University Press, 2009.

20 参见 ISO 8373: 2012. https://www.iso.org/obp/ui/#iso:std:iso:8373:ed-2:v1:en（访问时间 2018 年 4 月 30 日）

21 这个整合性哲学伦理框架最早由德国哲学家 Ludwig Siep 针对当代生命伦理学争论提出并发展。参考 L. Siep, Konkrete Ethik, 2. Auflage, Frankfurt a. Main, 2016 以及他的论文 "Zwei Formen der Ethik", in his collection *Moral und Gottesbild,* (Münster: Mentis, 2013, pp. 69 - 93。

22 参见 G. Simondon, *Du Mode d'existence des objets techniques*, Paris : Aubier, 1958 and B. Latour, "Where are the Missing Masses ? The Sociology of a Few Mundane Artefacts ", in the collection *Shaping Technology/Building Society*, ed. by Wiebe E. Bijker and John Law, Cambridge: The MIT Press, 1992, pp. 225 - 258。

23 *A Roadmap for US Robotics: From Internet to Robotics*, 2016 Edition, p. 23.

24 这里采纳的"能力路径"（capabilities approach）最早由美国当代经济学、诺贝尔经济学奖获得者 Amartya Sen 所提出。美国哲学家 Martha Nussbaum 借鉴古希腊亚里士多德伦理传统对于该能力路径进行发展。在此，我们则通过中国儒家伦理传统来发展"能力路径"。参考 A.

org/downloads/press/Executive_Summary_WR_Service_Robots_2017_1. pdf（visited on April 30,2018）

17 G. Veruggio and F. Operto, "Roboethics: Social and Ethical Implications of Robotics", in the *Springer Handbook of Robotics,* ed. by Bruno Siciliano and Oussama Khatib, Berlin: Springer, 2008, p. 1504.

18 K. Abney, "Robotics, Ethical Theory, and Metaethics: A Guide for the Perplexed", in the collection *Robot Ethics: The Ethical and Social Implications of Robotics,* ed. by Patrick Lin, Keith Abney and George A. Bekey, Cambridge: The MIT Press, 2012, p. 35.

19 W. Wallach and C. Allen, *Moral Machines: Teaching Robots Right from Wrong,* Oxford: Oxford University Press, 2009.

20 Cf. ISO 8373: 2012. https://www.iso.org/obp/ui/#iso:std:iso:8373:ed-2:v1:en （visited on April 30,2018）

21 Ethical holism is first proposed and developed by the contemporary German philosopher Ludwig Siep, especially with regards to the area of bioethics. Cf. Ludwig Siep, *Konkrete Ethik,* 2. Auflage, Frankfurt a. Main, 2016 and "Zwei Formen der Ethik", in his collection *Moral und Gottesbild,* Münster: Mentis, 2013, pp. 69 - 93.

22 Cf. G. Simondon, *Du Mode d'existence des objets techniques,* Paris : Aubier, 1958 and B. Latour, "Where are the Missing Masses? The Sociology of a Few Mundane Artefacts ", in the collection *Shaping Technology/Building Society,* ed. by Wiebe E. Bijker and John Law, Cambridge: The MIT Press, 1992, pp. 225 - 258.

23 *A Roadmap for US Robotics: From Internet to Robotics,* 2016 Edition, p. 23.

24 The capabilities approach is proposed and developed by contemporary American economist and philosopher, Amartya Sen and philosopher Martha Nussbaum, Cf. Amartya Sen, *Commodities and Capabilities,*

Sen, *Commodities and Capabilities,* New Delhi: Oxford University Press, 1999 and M. Nussbaum, *Creating Capabilities*, Cambridge: The Belknap Press of Harvard University Press, 2011。

[25] *A Roadmap for US Robotics: From Internet to Robotics,* 2016 Edition, p. 5.

[26] 参考 L. Siep, "Grundlagen der Konkreten Ethik", in his *Moral und Gottesbild, op. cit.,* pp. 174 - 176。

[27] 对于机器人引发伦理问题的广泛描述，参考 S. di Robotica and G. Veruggio, *EURON Roboethics Roadmap,* 2006, BS 8611: 2016. *Robots and Robotic Devices: Guide to the Ethical Design and Application of Robots and Robotic Systems,* 以及 The IEEE Global Initiative on Ethics of Autonomous and Intelligent Systems, *Ethically Alinged Design: A Vision for Prioritizing Human Well-being with Autonomous and Intelligent Systems,* Version 2. IEEE, 2017. http://standards.ieee.org/develop/indconn/ec/autonomous_systems.html（访问时间 2018 年 4 月 30 日）

[28] 参考 S. di Robotica and G. Veruggio, *EURON Roboethics Roadmap,* 2006, BS 8611: 2016. *Robots and Robotic Devices: Guide to the Ethical Design and Application of Robots and Robotic Systems,* 以及 The IEEE Global Initiative on Ethics of Autonomous and Intelligent Systems, *Ethically Alinged Design: A Vision for Prioritizing Human Well-being with Autonomous and Intelligent Systems,* Version 2. IEEE, 2017. http://standards.ieee.org/develop/indconn/ec/autonomous_systems.html（访问时间 2018 年 4 月 30 日）

[29] 参考中华人民共和国国务院新闻办公室 2017 年 12 月 15 日发布的《中国人权法治化保障的新进展》前瞻。http://www.scio.gov.cn/zfbps/ndhf/36088/Document/1613510/1613510.htm（访问时间 2018 年 4 月 30 日）

[30] 参考 P. Brey, "From Moral Agents to Moral Factors: The Structural Ethics Approach", in the collection *The Moral Status of Technical Artefacts,* ed. by Peter Kroes and Peter-Paul Verbeek, Dordrecht: Springer, 2014, pp. 125 - 142。

New Delhi: Oxford University Press, 1999 and Martha Nussbaum, *Creating Capabilities,* Cambridge: The Belknap Press of Harvard University Press, 2011.

[25] *A Roadmap for US Robotics: From Internet to Robotics,* 2016 Edition, p. 5.

[26] Cf. L. Siep. "Grundlagen der Konkreten Ethik", in his *Moral und Gottesbild, op. cit.,* pp. 174 - 176.

[27] Cf. S. d. Robotica and G. Veruggio, *EURON Roboethics Roadmap,* 2006, BS 8611: 2016. *Robots and Robotic Devices: Guide to the Ethical Design and Application of Robots and Robotic Systems,* and The IEEE Global Initiative on Ethics of Autonomous and Intelligent Systems. *Ethically Alinged Design: A Vision for Prioritizing Human Well-being with Autonomous and Intelligent Systems,* Version 2. IEEE, 2017. http://standards.ieee.org/develop/indconn/ec/autonomous_systems.html（visited on April 30,2018）

[28] Cf. S. d. Robotica and G. Veruggio, *EURON Roboethics Roadmap,* 2006, BS 8611: 2016. *Robots and Robotic Devices: Guide to the Ethical Design and Application of Robots and Robotic Systems,* and The IEEE Global Initiative on Ethics of Autonomous and Intelligent Systems, *Ethically Alinged Design: A Vision for Prioritizing Human Well-being with Autonomous and Intelligent Systems,* Version 2. IEEE, 2017. http://standards.ieee.org/develop/indconn/ec/autonomous_systems.html（visited on April 30,2018）

[29] The State Council Information Office of the People's Republic of China, *White Paper: New Progress in the Legal Protection of Human Rights in China,* 2017. http://www.scio.gov.cn/zfbps/ndhf/36088/Document/1613605/1613605.htm（visited on April 30,2018）

[30] Cf. P. A. E. Brey, "From Moral Agents to Moral Factors: The Structural Ethics Approach", in the collection *The Moral Status of Technical Artefacts,* ed. by Peter Kroes and Peter-Paul Verbeek, Dordrecht: Springer, 2014, pp. 125 - 142.

[31] 参考 S. di Robotica and G. Veruggio, *EURON Roboethics Roadmap,* 2006, BS 8611: 2016. *Robots and Robotic Devices: Guide to the Ethical Design and Application of Robots and Robotic Systems,* and The IEEE Global Initiative on Ethics of Autonomous and Intelligent Systems. *Ethically Alinged Design: A Vision for Prioritizing Human Well-being with Autonomous and Intelligent Systems,* Version 2. IEEE, 2017. http://standards.ieee.org/develop/indconn/ec/autonomous_systems.html（访问时间 2018 年 4 月 30 日）

[32] 参考 S. di Robotica and G. Veruggio, *EURON Roboethics Roadmap,* 2006, BS 8611: 2016. *Robots and Robotic Devices: Guide to the Ethical Design and Application of Robots and Robotic Systems,* and The IEEE Global Initiative on Ethics of Autonomous and Intelligent Systems. *Ethically Alinged Design: A Vision for Prioritizing Human Well-being with Autonomous and Intelligent Systems,* Version 2. IEEE, 2017. http://standards.ieee.org/develop/indconn/ec/autonomous_systems.html（访问时间 2018 年 4 月 30 日）

[33] 参考 S. di Robotica and G. Veruggio, *EURON Roboethics Roadmap,* 2006, BS 8611: 2016. *Robots and Robotic Devices: Guide to the Ethical Design and Application of Robots and Robotic Systems,* and The IEEE Global Initiative on Ethics of Autonomous and Intelligent Systems. *Ethically Alinged Design: A Vision for Prioritizing Human Well-being with Autonomous and Intelligent Systems,* Version 2. IEEE, 2017. http://standards.ieee.org/develop/indconn/ec/autonomous_systems.html（访问时间 2018 年 4 月 30 日）

[34] 参考 S. di Robotica and G. Veruggio, *EURON Roboethics Roadmap,* 2006, BS 8611: 2016. *Robots and Robotic Devices: Guide to the Ethical Design and Application of Robots and Robotic Systems,* and The IEEE Global Initiative on Ethics of Autonomous and Intelligent Systems, *Ethically Alinged Design: A Vision for Prioritizing Human Well-being with Autonomous and Intelligent Systems,* Version 2. IEEE, 2017. http://standards.ieee.org/develop/indconn/

[31] Cf. S. d. Robotica and G. Veruggio, *EURON Roboethics Roadmap,* 2006, BS 8611: 2016. *Robots and Robotic Devices: Guide to the Ethical Design and Application of Robots and Robotic Systems,* and The IEEE Global Initiative on Ethics of Autonomous and Intelligent Systems. *Ethically Alinged Design: A Vision for Prioritizing Human Well-being with Autonomous and Intelligent Systems,* Version 2. IEEE, 2017. http://standards.ieee.org/develop/indconn/ec/autonomous_systems.html（visited on April 30,2018）

[32] Cf. S. d. Robotica and G. Veruggio, *EURON Roboethics Roadmap,* 2006, BS 8611: 2016. *Robots and Robotic Devices: Guide to the Ethical Design and Application of Robots and Robotic Systems,* and The IEEE Global Initiative on Ethics of Autonomous and Intelligent Systems. *Ethically Alinged Design: A Vision for Prioritizing Human Well-being with Autonomous and Intelligent Systems,* Version 2. IEEE, 2017. http://standards.ieee.org/develop/indconn/ec/autonomous_systems.html（visited on April 30,2018）

[33] Cf. S. d. Robotica and G. Veruggio, *EURON Roboethics Roadmap,* 2006, BS 8611: 2016. *Robots and Robotic Devices: Guide to the Ethical Design and Application of Robots and Robotic Systems,* and The IEEE Global Initiative on Ethics of Autonomous and Intelligent Systems, *Ethically Alinged Design: A Vision for Prioritizing Human Well-being with Autonomous and Intelligent Systems,* Version 2. IEEE, 2017. http://standards.ieee.org/develop/indconn/ec/autonomous_systems.html（visited on April 30,2018）

[34] Cf. S. d. Robotica and G. Veruggio, *EURON Roboethics Roadmap,* 2006, BS 8611: 2016. *Robots and Robotic Devices: Guide to the Ethical Design and Application of Robots and Robotic Systems,* and The IEEE Global Initiative on Ethics of Autonomous and Intelligent Systems, *Ethically Alinged Design: A Vision for Prioritizing Human Well-being with Autonomous and Intelligent Systems,* Version 2. IEEE, 2017. http://standards.ieee.org/develop/indconn/

ec/autonomous_systems.html（访问时间 2018 年 4 月 30 日）

35 参考 M. Nussbaum and A. Sen (eds.), *The Quality of Life,* Oxford: Clarendon Press, 1993。

36 参考 The IEEE Global Initiative on Ethics of Autonomous and Intelligent Systems, *Ethically Alinged Design: A Vision for Prioritizing Human Well-being with Autonomous and Intelligent Systems,* Version 2. IEEE, 2017. http://standards.ieee.org/develop/indconn/ec/autonomous_systems.html（访问时间 2018 年 4 月 30 日）

37 参考 P. Brey, "From Moral Agents to Moral Factors: The Structural Ethics Approach", in the collection *The Moral Status of Technical Artefacts,* ed. by Peter Kroes and Peter-Paul Verbeek, Dordrecht: Springer, 2014, pp. 125 - 142, 以及 Aimee Van Wynsberghe, *Healthcare Robots,* Farnham: Ashgate, 2015。

38 例如，在老年人护理机器人的设计中，既需要考虑预防生理－心理过度依赖风险，也需要满足个体关怀的要求。机器人的设计和研发需要考虑这两个要求的设计平衡。相关讨论参见第五章。

39 例如，在个体价值优先的社群，机器人设计是否可以加大使用者能力强化的设计权重。

40 例如，BSI 颁布的标准指南 *Robots and Robotic Devices: Guide to the Ethical Design and Application of Robots and Robotic Systems* 和 IEEE 公布的国际行业性文件草案 *Ethically Aligned Design: A Vision for Prioritizing Human Well-Being with Autonomous and Intelligent Systems* 均采用相同的评估范畴进行机器人伦理评估。

41 A. Van Wynsberghe, *Healthcare Robots: Ethics, Design and Implementation,* Farnham: Ashgate, 2015.

42 A. Van Wynsberghe, *Healthcare Robots: Ethics, Design and Implementation,* Farnham: Ashgate, 2015, p.62.

43 J. Tronto, "Creating Caring Institutions: Politics, Plurality, and Purpose", in

ec/autonomous_systems.html (visited on April 30,2018)

[35] Cf. M. Nussbaum and A. Sen (eds.), *The Quality of Life,* Oxford: Clarendon Press, 1993.

[36] The IEEE Global Initiative on Ethics of Autonomous and Intelligent Systems, *Ethically Aligned Design: A Vision for Prioritizing Human Well-being with Autonomous and Intelligent Systems,* Version 2. IEEE, 2017. http://standards.ieee.org/develop/indconn/ec/autonomous_systems.html. (visited on April 30,2018)

[37] Cf. P. A. E. Brey, "From Moral Agents to Moral Factors: The Structural Ethics Approach", in the collection *The Moral Status of Technical Artefacts,* ed. by Peter Kroes and Peter-Paul Verbeek, Dordrecht: Springer, 2014, pp. 125 - 142 and Aimee Van Wynsberghe, *Healthcare Robots,* Farnham: Ashgate, 2015.

[38] For example, in the design of elder-caring robots, it is necessary to consider both the reduction of the risk of physio-psychological over-dependence and the need for sufficient amount of the individual caring. The design and development of this kind of robots need to take both requirements in a well-balanced way. For related, detailed discussion, see Chapter 5.

[39] For example, in a community that values individual excellence most, whether it is more permissible for robot designer to take the enhancement of the user's abilities as the only priority.

[40] For example, see *Robots and Robotic Devices: Guide to the Ethical Design and Application of Robots and Robotic Systems* issued by BSI, and *Ethically Aligned Design: A Vision for Prioritizing Human Well-Being with Autonomous and Intelligent Systems* issued by the IEEE, for the similar set of categories.

[41] A. Van Wynsberghe, *Healthcare Robots: Ethics, Design and Implementation,* Farnham: Ashgate, 2015.

[42] A. Van Wynsberghe, *Healthcare Robots: Ethics, Design and Implementation.* Farnham: Ashgate, 2015, p.62.

Ethics and Social Welfare, 4(2), 2010, pp. 158-171.

44 P. W. Singer, *Wired for War,* New York: Penguin, 2009.

45 G. Bekey, "Current Trends in Robotics: Technology and Ethics", in *Robot Ethics: The Ethical and Social Implications of Robotics,* ed. by Patrick Lin, Keith Abney and George A. Bekey, Cambridge: The MIT Press, 2012, pp. 17-34.

46 L. Floridi and J. W. Sanders, "On the Morality of Artificial Agents", in *Minds and Machines 14 (3),* 2004, pp. 349-379.

47 W. Wallach and C. Allen, *Moral Machines: Teaching Robots Right from Wrong,* Oxford: Oxford University Press, 2010（中译本：温德尔·瓦拉赫、科林·艾伦：《道德机器：如何让机器人明辨是非》，王小红主译，北京：北京大学出版社，2017年）。

48 S. G. Tzafestas, *Roboethics: A Navigating Overview,* Dordrecht: Springer, 2016.

49 M. Coeckelbergh, C. Pop, R. S. Vanderborght, A. Peca, S. Pintea, D. D. David, B. Vanderborght, "A Survey of Expectations About the Role of Robots in Robot-Assisted Therapy for Children with ASD: Ethical Acceptability, Trust, Sociability, Appearance, and Attachment", in *Science and Engineering Ethics, 22(1),* pp. 47-65.

50 A. Elders, "Robot Friends for Autistic Children: Monopoly Money or Counterfeit Currency?", in *Robot Ethics 2.0 : From Autonomous Cars to Artificial Intelligence,* ed. by P. Lin, R. Jenkins, K. Abney, Oxford: Oxford University Press, 2017, pp. 113-126.

[43] J. Tronto, "Creating Caring Institutions: Politics, Plurality, and Purpose", in *Ethics and Social Welfare, 4(2),* 2010, pp. 158-171.

[44] P. W. Singer, *Wired for War.* New York: Penguin, 2009.

[45] G. Bekey, "Current Trends in Robotics: Technology and Ethics", in *Robot Ethics: The Ethical and Social Implications of Robotics,* ed. by Patrick Lin, Keith Abney and George A. Bekey, Cambridge: The MIT Press, 2012, pp. 17-34.

[46] L. Floridi and J. W. Sanders, "On the Morality of Artificial Agents", in *Minds and Machines 14 (3),* 2004, pp. 349-379.

[47] W. Wallach and C. Allen, *Moral Machines: Teaching Robots Right from Wrong,* Oxford: Oxford University Press, 2010.

[48] S. G. Tzafestas, *Roboethics: A Navigating Overview,* Dordrecht: Springer, 2016.

[49] M. Coeckelbergh, C. Pop, R. S. Vanderborght, A. Peca, S. Pintea, D. D. David, B. Vanderborght, "A Survey of Expectations About the Role of Robots in Robot-Assisted Therapy for Children with ASD: Ethical Acceptability, Trust, Sociability, Appearance, and Attachment", in *Science and Engineering Ethics, 22(1),* pp. 47-65.

[50] A. Elders, "Robot Friends for Autistic Children: Monopoly Money or Counterfeit Currency?" in *Robot Ethics 2.0 : From Autonomous Cars to Artificial Intelligence,* ed. by P. Lin, R. Jenkins, K. Abney, Oxford: Oxford University Press, 2017, pp. 113-126.

中国机器人伦理标准化前瞻委员会

工作组组长

刘　哲（北京大学）

刘连庆（中国科学院沈阳自动化所）

撰写组成员（按音序）

安乐哲（北京大学）　　郭　耀（北京大学）

李麒麟（北京大学）　　刘　哲（北京大学）

罗定生（北京大学）　　尚新建（北京大学）

王启宁（北京大学）　　王彦晶（北京大学）

吴天岳（北京大学）　　徐向东（浙江大学）

咨询委员会委员（按音序）

陈自富（上海交通大学）

丁　宁（香港中文大学[深圳]）

段伟文（中国社会科学研究院哲学研究所）

梁　波（中国科学院沈阳自动化研究所）

牟　昱（中国科学院沈阳自动化研究所）

Committee Members

Directors

Liu Zhe (Peking University)

Liu LianQing (ShenYang Institute of Automation, CAS)

Group Members

Guo Yao (Peking University)

Li Qilin (Peking University)

Liu Zhe (Peking University)

Luo Dingsheng (Peking University)

Roger Ames (Peking University)

Shang Xinjian (Peking University)

Wang Qining (Peking University)

Wang Yanjing (Peking University)

Wu Tianyue (Peking University)

Xu Xiangdong (Zhejiang University)

Advisory Committee

Chen Zifu (Shanghai Jiao Tong University)

Ding Ning (The Chinese University of Hong Kong, Shen Zhen)

Duan Weiwen (Institute of Philosophy, CASS)

Liang Bo (ShenYang Institute of Automation, CAS)

Mou Yu (ShenYang Institute of Automation, CAS)

曲道奎（新松机器人自动化股份有限公司）

宋晓刚（机器人产业联盟）

王大宁（达闼科技有限公司）

王　虹（中国科学院沈阳自动化研究所）

王会方（南京市特种设备安全监督检验研究院）

王　健（东北大学）

吴　蒙（中国家用电器研究院）

徐全平（国家标准化管理委员会）

尹作重（全国自动化系统与集成标准化技术委员会）

张亚晨（中国家用电器研究院）

郑军奇（上海电器科学研究院）

Qu Daokui (SIASUN Robot & Automation CO., Ltd)

Song Xiaogang (China Robot Industry Alliance)

Wang Daning (CloudMinds)

Wang Hong (ShenYang Institute of Automation, CAS)

Wang Huifang (Nanjing Special Equipment Inspection Institute)

Wang Jian (Northeastern University)

Wu Meng (China Household Electric Appliance Research Institute)

Xu Quanping (Standardization Administration of the People's Republic of China)

Yin Zuozhong (China National Technical Committee for Automation Systems and Integration Standardization)

Zhang Yachen (China Household Electric Appliance Research Institute)

Zheng Junqi (Shanghai Electrical Apparatus Research Institute)

更正说明

本书的著作方式为"编著",而非"著",特此说明。